Design-for-Test for Digital IC's and Embedded Core Systems

Alfred L. Crouch

Prentice Hall PTR
Upper Saddle River, NJ 07458
www.phptr.com

Library of Congress Cataloging-in-Publication Data

```
Crouch, Alfred L.
     Design-for-test for digital IC's and embedded core systems /
Alfred L. Crouch.
        p.   cm.
     Includes bibliographical references and index.
     ISBN 0-13-084827-1
        1. Digital integrated circuits--Design and construction.
2. Electronic circuit design. 3. Automatic checkout equipment.
4. Embedded computer systems--Design and construction.  I. Title
TK7874.65.C76  1999
621.3815--dc21                                            99-23871
                                                              CIP
```

Editorial/production supervision: *Camille Trentacoste*
Acquisitions editor: *Bernard M. Goodwin*
Marketing manager: *Lisa Konzelman*n
Manufacturing manager: *Alan Fischer*
Cover design director: *Jerry Votta*
Cover designer: *Talar Agasyan*
Cover art: *Alamini designs*

© 1999 by Prentice Hall PTR
Prentice-Hall, Inc.
Upper Saddle River, New Jersey 07458

Prentice Hall books are widely used by corporations and government agencies for training, marketing, and resale. The publisher offers discounts on this book when ordered in bulk quantities. For more information, contact: Corporate Sales Department, Phone: 800-382-3419; Fax: 201-236-7141; Email corpsales@prenhall.com; or write: Prentice Hall PTR, Corporate Sales Department, One Lake Street, Upper Saddle River, NJ 07458.

All rights reserved. No part of this publication may be reproduced, in any form or by any means, without permission in writing from the publisher.

The use of the slides contained on the enclosed CD-ROM for personal study, design presentations, management presentations, engineering discussions, and internal training presentations is encouraged; however, these slides can't be used as the basis for a DFT for-profit class without the express written permission of the publisher.

All product or services mentioned in this book are the trademarks or service marks of their respective companies or organizations.

Reprinted with corrections March, 2000.

Printed in the United States of America
10 9 8 7 6 5 4 3

ISBN 0-13-084827-1

Prentice-Hall International (UK) Limited, *London*
Prentice-Hall of Australia Pty. Limited, *Sydney*
Prentice-Hall Canada Inc., *Toronto*
Prentice-Hall Hispanoamericana, S.A., *Mexico*
Prentice-Hall of India Private Limited, *New Delhi*
Prentice-Hall of Japan, Inc., *Tokyo*
Prentice-Hall of Asia Pte. Ltd., *Singapore*
Editora Prentice-Hall do Brasil, Ltda., *Rio de Janeiro*

Preface

This book is made primarily for design engineers and managers, and for test and design-for-test engineers and managers. It can also be used for students of digital design and test, as well. The purpose of this book is to introduce the basic concepts of test and design-for-test (DFT), and to then address the application of these concepts with an eye toward the trade-offs of the engineering budgets (silicon area, operating frequency target, power consumption, etc.), the business drivers, and the cost factors.

Currently, some very good test and DFT texts are available. However, many of them are from an academic focus. In my years of being part of the integrated circuit design community, I have had to train many IC designers and junior DFT engineers in test and design-for-test. I have discovered that corporate education is remarkably different from academic education. A design engineer on a project, who must learn and apply DFT techniques, is learning them while being responsible for 60+ hours of other design tasks per week and while meeting regular design deadlines and milestones. In this environment, learning the DFT tasks and techniques is difficult with a book that focuses on the "mathematical" or "theoretical" point of view. History has taught me that a direct "how to do it" text is more effective.

Another interesting aspect of the competitive corporate environment is that the design process may be "factory-ized." The overall design process for a chip or a portion of a chip is no longer the responsibility of the design engineer, but of teams of chip design functions. For example, the logic gate cells may be designed and characterized by one group (standard cell and library development), and the design may be modeled and synthesized by a different group (HDL design and synthesis), verified by yet another group (formal and simulation verification), and ultimately, mapped to a physical process by yet another group (floorplanning, place&route, and physical design). In this case, the teaching of DFT techniques must be spread out to the various organizations contributing to the overall design. A teaching description of a DFT technique, such as scan design, is not effective if it is not related to the tasks, scheduling, trade-offs, and the separations into the various organizational elements. Again, history and experience have taught me that an effective text here is one that relates the topic being addressed to the design methodology and design flow.

So direct experience in corporate technical training and teaching has led to the style and content of this "practical guide" on the test and Design-for-Test (DFT) topics of scan test, embedded memory test, and embedded core test. This text has been developed more along the lines of a "just what you need to know—and how to do it" guide that explains the topic, explains the trade-offs, and relates the topic to the design flow. My hope is that using this text will reduce the "learning curve" involved in the application of test and design-for-test techniques, and will result in designs that have a higher quality-reliability level and a lower cost-of-test.

A practical text on DFT and DFT techniques, based on the industry point of view, is needed right now for several reasons. First, the "cost of test" is beginning to dominate the recurring (per-part) cost involved in the manufacturing of the final silicon product for many of the consumer markets—parts with high volume and a low selling price. Second, shorter product lifetimes and increased time-to-market (TTM) and time-to-volume (TTV) pressures are forcing the need to have some form of structured, repeatable and automatable test features included in the device as part of the overall design methodology. Third, the move to reuse cores, and core-based design, as a reaction to shrinking process geometries and TTM pressures, is also forcing designed-in test features to become portable since design units may be distributed and reused in several different chip designs with completely different configurations. And finally, the shrinking process geometries also enable "system-on-a-chip" and ULSI (Ultra-Large Scale Integrated) designs with massive integration—more integration means more faults and more vectors—which leads to a test data management and cost-of-test problems.

Taken all together, these changes in parts of the semiconductor design industry are changing the way test and DFT are viewed, addressed, and implemented. Organizations that once ignored DFT are now being dragged kicking and screaming into modern age because of test cost, TTM, TTV, test data volume, and having to deal with the test features delivered with commercially available cores. Test is one of the three major components of recurring per-part cost involved with the manufacture and selling of digital semiconductor integrated circuits (with the cost of silicon die and the cost of packaging being the other two). As with every product, trade-offs are made to achieve a quality level and to fit within a target cost profile. I hope that this text will eliminate the view that understanding the cost-of-test and applying DFT during the design phase of a product is a "black art" for organizations and individuals that must address managing the cost factors of a chip design.

If you have questions or comments, I can be contacted at *Al_Crouch@prodigy.net*.

Acknowledgments

A book like this could never have been completed without the support of many people. I would like to thank a couple of my early mentors for setting me on the course to becoming the design-for-test professional that I turned out to be, Greg Young and Andy Halliday. The knowledge and data gathered for many of the test techniques and trade-offs described in this book would not have been possible without the work being accomplished to implement these techniques in real chips. To this end, I would like to recognize the hard work, support, and commitment of the Motorola ColdFire® DFT team, and other DFT professionals that I work with and have worked with at Motorola: Renny Eisele, Teresa McLaurin, John Potter, Michael Mateja, Dat Tran, Jennifer McKeown, Jim Johnson, Matthew Pressley, Clark Shepard, Bill Underwood, and Jason Doege. I would like to thank Janusz Rajski and Pat Scoggin of Mentor Graphics for convincing me that this information would make a good book (and that I could write it). And I would also like to thank Grady Giles for all his years of invaluable guidance and assistance. I would also like to thank my reviewers, Elizabeth Rudnick of the University of Illinois, and Nur Touba of the University of Texas.

The person I need to thank the most, however, is my wife, Kelly, for allowing me to work my "second" job of book writing for so long.

ColdFire is a registered trademark of Motorola

*For my sister, Monika,
who works harder just
staying alive every day
than I can ever hope
to match.*

Contents

Preface   iii

Acknowledgments   v

Introduction   xxiii

1 Test and Design-for-Test Fundamentals   1
1.1   Introduction to Test and DFT Fundamentals   3
    1.1.1   Purpose   3
    1.1.2   Introduction to Test, the Test Process, and Design-for-Test   3
    1.1.3   Concurrent Test Engineering   4
1.2   The Reasons for Testing   7
    1.2.1   Why Test? Why Add Test Logic?   7
    1.2.2   Pro and Con Perceptions of DFT   7
1.3   The Definition of Testing   10
    1.3.1   What Is Testing?   10
    1.3.2   Stimulus   11
    1.3.3   Response   11
1.4   Test Measurement Criteria   13
    1.4.1   What Is Measured?   13
    1.4.2   Fault Metric Mathematics   14

1.5	Fault Modeling		16
	1.5.1	Physical Defects	16
	1.5.2	Fault Modeling	16
1.6	Types of Testing		20
	1.6.1	Functional Testing	20
	1.6.2	Structural Testing	20
	1.6.3	Combinational Exhaustive and Pseudo-Exhaustive Testing	20
	1.6.4	Full Exhaustive Testing	21
	1.6.5	Test Styles	21
1.7	Manufacturing Test		23
	1.7.1	The Manufacturing Test Process	23
	1.7.2	Manufacturing Test Load Board	23
	1.7.3	Manufacturing Test Program	23
1.8	Using Automatic Test Equipment		25
	1.8.1	Automatic Test Equipment	25
	1.8.2	ATE Limitations	25
	1.8.3	ATE Cost Considerations	25
1.9	Test and Pin Timing		27
	1.9.1	Tester and Device Pin Timing	27
	1.9.2	Tester Edge Sets	27
	1.9.3	Tester Precision and Accuracy	28
1.10	Manufacturing Test Program Components		30
	1.10.1	The Pieces and Parts of a Test Program	30
	1.10.2	Test Program Optimization	32
1.11	Recommended Reading		33

2 Automatic Test Pattern Generation Fundamentals   35

2.1	Introduction to Automatic Test Pattern Generation		37
	2.1.1	Purpose	37
	2.1.2	Introduction to Automated Test Pattern Generation	37
	2.1.3	The Vector Generation Process Flow	38
2.2	The Reasons for ATPG		41
	2.2.1	Why ATPG?	41
	2.2.2	Pro and Con Perceptions of ATPG	41
2.3	The Automatic Test Pattern Generation Process		44
	2.3.1	Introduction to ATPG	44

2.4	Introducing the Combinational Stuck-At Fault	47
2.4.1	Combinational Stuck-At Faults	47
2.4.2	Combinational Stuck-At Fault Detection	47
2.5	Introducing the Delay Fault	49
2.5.1	Delay Faults	49
2.5.2	Delay Fault Detection	49
2.6	Introducing the Current-Based Fault	52
2.6.1	Current-Based Testing	52
2.6.2	Current-Based Testing Detection	52
2.7	Testability and Fault Analysis Methods	54
2.7.1	Why Conduct ATPG Analysis or Testability Analysis?	54
2.7.2	What Types of Testability Analysis Are Available?	54
2.7.3	Fault Effective Circuits	54
2.7.4	Controllability-Observability Analysis	55
2.7.5	Circuit Learning	56
2.8	Fault Masking	58
2.8.1	Causes and Effects of Fault Masking	58
2.8.2	Fault Masking on Various Fault Models	58
2.9	Stuck Fault Equivalence	60
2.9.1	Fault Equivalence Optimization	60
2.9.2	Fault Equivalence Side Effects	60
2.10	Stuck-At ATPG	62
2.10.1	Fault Selection	62
2.10.2	Exercising the Fault	63
2.10.3	Detect Path Sensitization	63
2.11	Transition Delay Fault ATPG	65
2.11.1	Using ATPG with Transition Delay Faults	65
2.11.2	Transition Delay Is a Gross Delay Fault	66
2.12	Path Delay Fault ATPG	68
2.12.1	Path Delay ATPG	68
2.12.2	Robust Fault Detection	68
2.12.3	The Path Delay Design Description	69
2.12.4	Path Enumeration	69
2.13	Current-Based Fault ATPG	71
2.13.1	Current-Based ATPG Algorithms	71

2.14	Combinational versus Sequential ATPG	73
2.14.1	Multiple Cycle Sequential Test Pattern Generation	73
2.14.2	Multiple Time Frame Combinational ATPG	74
2.14.3	Two-Time-Frame ATPG Limitations	75
2.14.4	Cycle-Based ATPG Limitations	75
2.15	Vector Simulation	77
2.15.1	Fault Simulation	77
2.15.2	Simulation for Manufacturing Test	77
2.16	ATPG Vectors	80
2.16.1	Vector Formats	80
2.16.2	Vector Compaction and Compression	80
2.17	ATPG-Based Design Rules	83
2.17.1	The ATPG Tool "NO" Rules List	83
2.17.2	Exceptions to the Rules	84
2.18	Selecting an ATPG Tool	87
2.18.1	The Measurables	87
2.18.2	The ATPG Benchmark Process	88
2.19	ATPG Fundamentals Summary	91
2.19.1	Establishing an ATPG Methodology	91
2.20	Recommended Reading	92

3 Scan Architectures and Techniques   93

3.1	Introduction to Scan-Based Testing	95
3.1.1	Purpose	95
3.1.2	The Testing Problem	95
3.1.3	Scan Testing	96
3.1.4	Scan Testing Misconceptions	96
3.2	Functional Testing	99
3.3	The Scan Effective Circuit	101
3.4	The Mux-D Style Scan Flip-Flops	103
3.4.1	The Multiplexed-D Flip-Flop Scan Cell	103
3.4.2	Perceived Silicon Impact of the Mux-D Scan Flip-Flop	103
3.4.3	Other Types of Scan Flip-Flops	103
3.4.4	Mixing Scan Styles	104
3.5	Preferred Mux-D Scan Flip-Flops	106
3.5.1	Operation Priority of the Multiplexed-D Flip-Flop Scan Cell	106
3.5.2	The Mux-D Flip-Flop Family	106

3.6	The Scan Shift Register or Scan Chain	108
3.6.1	The Scan Architecture for Test	108
3.6.2	The Scan Shift Register (a.k.a The Scan Chain)	108
3.7	Scan Cell Operations	110
3.7.1	Scan Cell Transfer Functions	110
3.8	Scan Test Sequencing	112
3.9	Scan Test Timing	115
3.10	Safe Scan Shifting	118
3.11	Safe Scan Sampling: Contention-Free Vectors	120
3.11.1	Contention-Free Vectors	120
3.12	Partial Scan	122
3.12.1	Scan Testing with Partial-Scan	122
3.12.2	Sequential ATPG	122
3.13	Multiple Scan Chains	125
3.13.1	Advantages of Multiple Scan Chains	125
3.13.2	Balanced Scan Chains	125
3.14	The Borrowed Scan Interface	128
3.14.1	Setting up a Borrowed Scan Interface	128
3.14.2	The Shared Scan Input Interface	128
3.14.3	The Shared Scan Output Interface	129
3.15	Clocking, On-Chip Clock Sources, and Scan	131
3.15.1	On-Chip Clock Sources and Scan Testing	131
3.15.2	On-Chip Clocks and Being Scan Tested	131
3.16	Scan-Based Design Rules	134
3.16.1	Scan-Based DFT and Design Rules	134
3.16.2	The Rules	134
3.17	Stuck-At (DC) Scan Insertion	139
3.17.1	DC Scan Insertion	139
3.17.2	Extras	139
3.17.3	DC Scan Insertion and Multiple Clock Domains	140
3.18	Stuck-At Scan Diagnostics	142
3.18.1	Implementing Stuck-At Scan Diagnostics	142
3.18.2	Diagnostic Fault Simulation	142
3.18.3	Functional Scan-Out	143
3.19	At-Speed Scan (AC) Test Goals	145
3.19.1	AC Test Goals	145
3.19.2	Cost Drivers	145

3.20	At-Speed Scan Testing	148
	3.20.1 Uses of At-Speed Scan Testing	148
	3.20.2 At-Speed Scan Sequence	148
	3.20.3 At-Speed Scan versus DC Scan	148
3.21	The At-Speed Scan Architecture	150
	3.21.1 At-Speed Scan Interface	150
	3.21.2 At-Speed "Safe Shifting" Logic	150
	3.21.3 At-Speed Scan Sample Architecture	150
3.22	The At-Speed Scan Interface	152
	3.22.1 At-Speed Scan Shift Interface	152
	3.22.2 At-Speed Scan Sample Interface	152
3.23	Multiple Clock and Scan Domain Operation	154
	3.23.1 Multiple Timing Domains	154
3.24	Scan Insertion and Clock Skew	157
	3.24.1 Multiple Clock Domains, Clock Skew, and Scan Insertion	157
	3.24.2 Multiple Time Domain Scan Insertion	158
3.25	Scan Insertion for At-Speed Scan	161
	3.25.1 Scan Cell Substitution	161
	3.25.2 Scan Control Signal Insertion	161
	3.25.3 Scan Interface Insertion	161
	3.25.4 Other Considerations	161
3.26	Critical Paths for At-Speed Scan	163
	3.26.1 Critical Paths	163
	3.26.2 Critical Path Selection	163
	3.26.3 Path Filtering	164
	3.26.4 False Path Content	165
	3.26.5 Real Critical Paths	166
	3.26.6 Critical Path Scan-Based Diagnostics	166
3.27	Scan-Based Logic BIST	168
	3.27.1 Pseudo-Random Pattern Generation	168
	3.27.2 Signature Analysis	168
	3.27.3 Logic Built-In Self-Test	168
	3.27.4 LFSR Science (A Quick Tutorial)	169
	3.27.5 X-Management	170
	3.27.6 Aliasing	170
3.28	Scan Test Fundamentals Summary	173
3.29	Recommended Reading	174

4 Memory Test Architectures and Techniques — 175

- 4.1 Introduction to Memory Testing — 177
  - 4.1.1 Purpose — 177
  - 4.1.2 Introduction to Memory Test — 177
- 4.2 Types of Memories — 180
  - 4.2.1 Categorizing Memory Types — 180
- 4.3 Memory Organization — 183
  - 4.3.1 Types of Memory Organization — 183
- 4.4 Memory Design Concerns — 186
  - 4.4.1 Trade-Offs in Memory Design — 186
- 4.5 Memory Integration Concerns — 188
  - 4.5.1 Key Issues in Memory Integration — 188
- 4.6 Embedded Memory Testing Methods — 190
  - 4.6.1 Memory Test Methods and Options — 190
- 4.7 The Basic Memory Testing Model — 193
  - 4.7.1 Memory Testing — 193
  - 4.7.2 Memory Test Fault Model — 193
  - 4.7.3 Memory Test Failure Modes — 193
- 4.8 The Stuck-At Bit-Cell Based Fault Models — 195
  - 4.8.1 Stuck-At Based Memory Bit-Cell Fault Models — 195
  - 4.8.2 Stuck-At Fault Exercising and Detection — 195
- 4.9 The Bridging Defect-Based Fault Models — 197
  - 4.9.1 Bridging Defect-Based Memory Test Fault Models — 197
  - 4.9.2 Linking Defect Memory Test Fault Models — 197
  - 4.9.3 Bridging Fault Exercising and Detection — 197
- 4.10 The Decode Fault Model — 199
  - 4.10.1 Memory Decode Fault Models — 199
  - 4.10.2 Decode Fault Exercising and Detection — 199
- 4.11 The Data Retention Fault — 201
  - 4.11.1 Memory Test Data Retention Fault Models — 201
  - 4.11.2 DRAM Refresh Requirements — 201
- 4.12 Diagnostic Bit Mapping — 203
  - 4.12.1 Memory Test Diagnostics: Bit Mapping — 203
- 4.13 Algorithmic Test Generation — 205
  - 4.13.1 Introduction to Algorithmic Test Generation — 205
  - 4.13.2 Automatic Test Generation — 205
  - 4.13.3 BIST-Based Algorithmic Testing — 206

4.14	Memory Interaction with Scan Testing	208
	4.14.1 Scan Test Considerations	208
	4.14.2 Memory Interaction Methods	208
	4.14.3 Input Observation	208
	4.14.4 Output Control	208
4.15	Scan Test Memory Modeling	210
	4.15.1 Modeling the Memory for ATPG Purposes	210
	4.15.2 Limitations	210
4.16	Scan Test Memory Black-Boxing	212
	4.16.1 The Memory Black-Boxing Technique	212
	4.16.2 Limitations and Concerns	212
4.17	Scan Test Memory Transparency	214
	4.17.1 The Memory Transparency Technique	214
	4.17.2 Limitations and Concerns	214
4.18	Scan Test Memory Model of The Fake Word	216
	4.18.1 The Fake Word Technique	216
	4.18.2 Limitations and Concerns	216
4.19	Memory Test Requirements for MBIST	218
	4.19.1 Memory Test Organization	218
4.20	Memory Built-In Self-Test Requirements	220
	4.20.1 Overview of Memory BIST Requirements	220
	4.20.2 At-Speed Operation	220
4.21	An Example Memory BIST	222
	4.21.1 A Memory Built-In Self-Test	222
	4.21.2 Optional Operations	223
	4.21.3 An Example Memory Built-In Self-Test	223
4.22	MBIST Chip Integration Issues	225
	4.22.1 Integrating Memory BIST	225
4.23	MBIST Integration Concerns	227
	4.23.1 MBIST Default Operation	227
4.24	MBIST Power Concerns	229
	4.24.1 Banked Operation	229
4.25	MBIST Design—Using LFSRs	231
	4.25.1 Pseudo-Random Pattern Generation for Memory Testing	231
	4.25.2 Signature Analysis and Memory Testing	231
	4.25.3 Signature Analysis and Diagnostics	231

4.26	Shift-Based Memory BIST	234
	4.26.1 Shift-Based Memory Testing	234
	4.26.2 Output Assessment	234
4.27	ROM BIST	236
	4.27.1 Purpose and Function of ROM BIST	236
	4.27.2 The ROM BIST Algorithm	237
	4.27.3 ROM MISR Selection	237
	4.27.4 Signature Compare Method	238
4.28	Memory Test Summary	240
4.29	Recommended Reading	240

5 Embedded Core Test Fundamentals — 241

5.1	Introduction to Embedded Core Testing	243
	5.1.1 Purpose	243
	5.1.2 Introduction to Embedded Core-Based Chip Testing	243
	5.1.3 Reuse Cores	244
	5.1.4 Chip Assembly Using Reuse Cores	244
5.2	What Is a Core?	246
	5.2.1 Defining Cores	246
	5.2.2 The Core DFT and Test Problem	246
	5.2.3 Built-In DFT	246
5.3	What is Core-Based Design?	248
	5.3.1 Design of a Core-Based Chip	248
	5.3.2 Core-Based Design Fundamentals	248
5.4	Reuse Core Deliverables	250
	5.4.1 Embedded Core Deliverables	250
5.5	Core DFT Issues	252
	5.5.1 Embedded Core-Based Design Test Issues	252
5.6	Development of a ReUsable Core	256
	5.6.1 Embedded Core Considerations for DFT	256
5.7	DFT Interface Considerations—Test Signals	262
	5.7.1 Embedded Core Interface Considerations for DFT—Test Signals	262
5.8	Core DFT Interface Concerns—Test Access	265
	5.8.1 Test Access to the Core Interface	265

5.9	DFT Interface Concerns—Test Wrappers	268
5.9.1	The Test Wrapper as a Signal reduction Element	268
5.9.2	The Test Wrapper as a Frequency Interface	268
5.9.3	The Test Wrapper as a Virtual Test Socket	269
5.10	The Registered Isolation Test Wrapper	271
5.11	The Slice Isolation Test Wrapper	273
5.12	The Isolation Test Wrapper—Slice Cell	275
5.13	The Isolation Test Wrapper—Core DFT Interface	277
5.14	Core Test Mode Default Values	279
5.14.1	Internal versus External Test Quiescence Defaults Application	279
5.15	DFT Interface Wrapper Concerns	281
5.15.1	Lack of Bidirectional Signals	281
5.15.2	Test Clock Source Considerations	281
5.16	DFT Interface Concerns—Test Frequency	284
5.16.1	Embedded Core Interface Concerns for DFT—Test Frequency	284
5.16.2	Solving the Frequency Problem	284
5.17	Core DFT Development	286
5.17.1	Internal Parallel Scan	286
5.17.2	Wrapper Parallel Scan	286
5.17.3	Embedded Memory BIST	287
5.17.4	Other DFT Features	287
5.18	Core Test Economics	289
5.18.1	Core DFT, Vectors, and Test Economics	289
5.18.2	Core Selection with Consideration to DFT Economics	289
5.19	Chip Design with a Core	292
5.19.1	Elements of a Core-Based Chip	292
5.19.2	Embedded Core Integration Concerns	292
5.19.3	Chip-Level DFT	293
5.20	Scan Testing the Isolated Core	296
5.21	Scan Testing the Non-Core Logic	298
5.21.1	Scan Testing the Non-Core Logic in Isolation	298
5.21.2	Chip-Level Testing and Tester Edge Sets	298
5.22	User Defined Logic Chip-Level DFT Concerns	300
5.23	Memory Testing with BIST	302

5.24	Chip-Level DFT Integration Requirements	304
	5.24.1 Embedded Core-Based DFT Integration Architecture	304
	5.24.2 Physical Concerns	305
5.25	Embedded Test Programs	307
5.26	Selecting or Receiving a Core	309
5.27	Embedded Core DFT Summary	311
5.28	Recommended Reading	311

About the CD   313

Glossary of Terms   317

Index   341

About the Author   349

Figures

1 Test and Design-for-Test Fundamentals — 1

Figure 1-1	Cost of Product	2
Figure 1-2	Concurrent Test Engineering	5
Figure 1-3	Why Test?	6
Figure 1-4	Definition of Testing	9
Figure 1-5	Measurement Criteria	12
Figure 1-6	Fault Modeling	15
Figure 1-7	Types of Testing	19
Figure 1-8	Manufacturing Test Load Board	22
Figure 1-9	Using ATE	24
Figure 1-10	Pin Timing	26
Figure 1-11	Test Program Components	29

2 Automatic Test Pattern Generation Fundamentals — 35

Figure 2-1	The Overall Pattern Generation Process	36
Figure 2-2	Why ATPG?	40
Figure 2-3	The ATPG Process	43
Figure 2-4	Combinational Stuck-At Fault	46
Figure 2-5	The Delay Fault	48
Figure 2-6	The Current Fault	51
Figure 2-7	Stuck-at Fault Effective Circuit	53
Figure 2-8	Fault Masking	57

Figure 2-9	Fault Equivalence Example	59
Figure 2-10	Stuck-At Fault ATPG	61
Figure 2-11	Transition Delay Fault ATPG	64
Figure 2-12	Path Delay Fault ATPG	67
Figure 2-13	Current Fault ATPG	70
Figure 2-14	Two-Time Frame ATPG	72
Figure 2-15	Fault Simulation Example	76
Figure 2-16	Vector Compression and Compaction	79
Figure 2-17	Some Example Design Rules for ATPG Support	82
Figure 2-18	ATPG Measurables	86

3 Scan Architectures and Techniques — 93

Figure 3-1	Introduction to Scan-Based Testing	94
Figure 3-2	An Example Non-Scan Circuit	98
Figure 3-3	Scan Effective Circuit	100
Figure 3-4	Flip-Flop versus Scan Flip-Flop	102
Figure 3-5	Example Set-Scan Flip-Flops	105
Figure 3-6	An Example Scan Circuit with a Scan Chain	107
Figure 3-7	Scan Element Operations	109
Figure 3-8	Example Scan Test Sequencing	111
Figure 3-9	Example Scan Testing Timing	114
Figure 3-10	Safe Scan Shifting	117
Figure 3-11	Safe Scan Vectors	119
Figure 3-12	Partial Scan	121
Figure 3-13	Multiple Scan Chains	124
Figure 3-14	The Borrowed Scan Interface	127
Figure 3-15	Clocking and Scan	130
Figure 3-16	Scan-Based Design Rules	133
Figure 3-17	DC Scan Insertion	138
Figure 3-18	Stuck-At Scan Diagnostics	141
Figure 3-19	At-Speed Scan Goals	144
Figure 3-20	At-Speed Scan Testing	147
Figure 3-21	At-Speed Scan Architecture	149
Figure 3-22	At-Speed Scan Interface	151
Figure 3-23	Multiple Scan and Timing Domains	153
Figure 3-24	Clock Skew and Scan Insertion	156
Figure 3-25	Scan Insertion for At-Speed Scan	160

Figures   xxi

Figure 3-26	Critical Paths for At-Speed Testing	162
Figure 3-27	Logic BIST	167
Figure 3-28	Scan Test Fundamentals Summary	172

4 Memory Test Architectures and Techniques   175

Figure 4-1	Introduction to Memory Testing	176
Figure 4-2	Memory Types	179
Figure 4-3	Simple Memory Organization	182
Figure 4-4	Memory Design Concerns	185
Figure 4-5	Memory Integration Concerns	187
Figure 4-6	Embedded Memory Test Methods	189
Figure 4-7	Simple Memory Model	192
Figure 4-8	Bit-Cell and Array Stuck-At Faults	194
Figure 4-9	Array Bridging Faults	196
Figure 4-10	Decode Faults	198
Figure 4-11	Data Retention Faults	200
Figure 4-12	Memory Bit Mapping	202
Figure 4-13	Algorithmic Test Generation	204
Figure 4-14	Scan Boundaries	207
Figure 4-15	Memory Modeling	209
Figure 4-16	Black Box Boundaries	211
Figure 4-17	Memory Transparency	213
Figure 4-18	The Fake Word Technique	215
Figure 4-19	Memory Test Needs	217
Figure 4-20	Memory BIST Requirements	219
Figure 4-21	An Example Memory BIST	221
Figure 4-22	MBIST Integration Issues	224
Figure 4-23	MBIST Default Values	226
Figure 4-24	Banked Operation	228
Figure 4-25	LFSR-Based Memory BIST	230
Figure 4-26	Shift-Based Memory BIST	233
Figure 4-27	ROM BIST	235
Figure 4-28	Memory Test Summary	239

5 Embedded Core Test Fundamentals   241

| Figure 5-1 | Introduction to Embedded Core Test and Test Integration | 242 |
| Figure 5-2 | What Is a Core? | 245 |

Figure 5-3	Chip Designed with Core	247
Figure 5-4	Reuse Core Deliverables	249
Figure 5-5	Core DFT Issues	251
Figure 5-6	Core Development DFT Considerations	255
Figure 5-7	DFT Core Interface Considerations	261
Figure 5-8	DFT Core Interface Concerns	264
Figure 5-9	DFT Core Interface Considerations	267
Figure 5-10	Registered Isolation Test Wrapper	270
Figure 5-11	Slice Isolation Test Wrapper	272
Figure 5-12	Slice Isolation Test Wrapper Cell	274
Figure 5-13	Core DFT Connections through the Test Wrapper	276
Figure 5-14	Core DFT Connections with Test Mode Gating	278
Figure 5-15	Other Core Interface Signal Concerns	280
Figure 5-16	DFT Core Interface Frequency Considerations	283
Figure 5-17	A Reuse Embedded Core's DFT Features	285
Figure 5-18	Core Test Economics	288
Figure 5-19	Chip with Core Test Architecture	291
Figure 5-20	Isolated Scan-Based Core-Testing	295
Figure 5-21	Scan Testing the Non-Core Logic	297
Figure 5-22	UDL Chip-Level DFT Concerns	299
Figure 5-23	Memory Testing the Device	301
Figure 5-24	DFT Integration Architecture	303
Figure 5-25	Test Program Components	306
Figure 5-26	Selecting or Receiving a Core	308
Figure 5-27	Embedded Core DFT Summary	310

Introduction

There is a lot of diverse information in this book, and although I would love to have everyone voraciously devour the book from the front cover to the back cover, many readers will be interested in reading only the pieces and parts that pertain to their needs. Since the practical application of design-for-test is a very interleaved and interrelated subject, none of the test and design-for-test techniques can exist in a total vacuum without some consideration for the other related test topics. For example, the application of a scan test architecture to a simple device with logic and memory, requires that some consideration be given to the interaction of the scan architecture with the memory array and the memory test architecture. So the study of scan will also require the study of some memory techniques. I have resisted the temptation to repeat information in several places, I have created a guide of sorts to the use of the book. This means that an individual who is interested only in applying scan test techniques to a chip-level design does need to read information in several sections to complete a course of study on the topic of scan. I will attempt to provide that type of outline here.

Also, I wish to apologize right here for writing in much the same manner as I speak. I have a habit of making my point by using slang and jargon. Since this book is based on the "corporate" or "business" point of view, the language I use is the language I learned at work, which may not line up exactly with the prescribed academic terms. To be completely honest, I learned most of what is in this book by the seat of my pants, applying DFT techniques to designs while my hair was on fire from schedule pressure—I did not spend a lot of time reading academic texts to learn my art (or science, as the case may be).

To begin with, the book contains five chapters, which can be thought of from a content point of view as lining up with the following main headings: Test, ATPG, Scan, Memory Test, and Cores. A quick synopsis of each chapter follows:

Chapter 1: Test and Design-for-Test Fundamentals: this chapter contains the terms, definitions, and the information involved with test to provide a basic understanding of just what test and design-for-test (DFT) are, how they are done, why they are done, what is being tested, what test is measuring and accomplishing, what equipment is used for test, and what the engineering and cost trade-off drivers are. It is a very basic chapter that can be used by a novice or a junior test or DFT professional to come up to speed on the requirements, actions, and language of test.

Chapter 2: Automatic Test Pattern Generation Fundamentals: this chapter describes the process of automating the onerous task of vector generation and reducing the time-to-volume by supporting an automatic test pattern generation (ATPG) methodology. Some of the analyses and techniques used in vector generation for both AC (dynamic) and DC (static) fault models are described to give an understanding of why certain rules must be applied to the hardware design and of how to reduce the size of the vector set or the time involved in generating the vector set. Also discussed are the measurables and trade-offs involved with the ATPG process so that an evaluation, or benchmark comparison, can be done between various ATPG tools to assist in the selection of the methodology and tool that is right for the application.

Chapter 3: Scan Architectures and Techniques: this chapter is about the scan test methodology and begins with the fundamentals of scan design and operation, and the design concerns and trade-offs involved with adopting a scan design. Also included are some techniques on the installation of scan into a design, how to deal with common problems such as safe-shifting, contention-free vectors, shift timing, and clock skew. Finally, some information is included on reducing test time by shifting the scan architecture at the rated functional frequency (at-speed scan), and on using scan to test for AC goals by operating the scan sample at the rated frequency (AC scan) based on using critical paths extracted from a design's timing analysis.

Chapter 4: Memory Test Architectures and Techniques: this chapter is about memory testing, memory interaction with the scan test architecture, and the adoption of memory built-in self-test (MBIST). This chapter begins with the fundamentals of memory test, and expands into the architectures involved with the coexistence with scan, and eventually describes the architectures and integration of using built-in self-test. The delivery of memories as cores with BIST (BIST'ed memories) includes information on how to deal with the integration of large numbers of memory cores, and how to minimize the routing problem, the power problem, the extraction of characterization or debug data problem, and the coordinated data retention problem.

Chapter 5: Embedded Core Test Fundamentals: this chapter is about creating testable designs with testable and accessible embedded cores. This chapter begins with the terms, definitions, issues, and trade-offs involved with the new style of device design known as "embedded IP" (intellectual property), embedded core, or core-based design. The embedded core design process is addressed in two main aspects, creating a testable reuse core, and implementing a chip design with embedded reusable core products. This chapter relies greatly on understanding the information from chapters 1, 3, and 4.

Guide to the Book

At first glance, this book may seem to be formatted a bit strangely. The beginning of each main subject heading within a chapter is a graphic that is then followed by several sub-sections of descriptive and teaching text. The graphics are the slides that I use when I teach this information as a course (all the graphics are available in full color with the included CD-ROM). You will notice that sometimes the graphic and the text are interrelated and the graphic supports the text; however, sometimes the graphic does not seem to be directly linked to the text, and sometimes the graphic will be an alternate description of the text. This is the nature of presentation material—the graphic must not be so complicated and busy that it confuses the viewer during a presentation or class—however, in the context of a reference book, readers sometimes enjoy a complicated graphic to trace through and analyze as they are learning the information. Since this text was designed to be an introductory and practical guide to the subject of Test and DFT, I decided to leave the presentation material as the graphics instead of replacing them with complicated diagrams.

The flow of information in this text also derives from the way I teach my courses in DFT. When I teach an entire comprehensive class, I naturally start with the basic information and language of test and then I move on to test pattern generation. With these two subjects as a basic foundation, I then move on to scan and then memory test, and I use scan and memory testing as a foundation to begin teaching core test. However, sometimes I am asked to teach just ATPG, just scan, just memory testing, or just core testing.

I could never hope to categorize every possible iteration of use for this book (and I don't mean its use as a doorstop or paperweight). However, I have taught several different courses of study with the information contained within this book, and to this end, I will describe what I think are the most common courses of study and outline the "paths" that must be taken through the book. These courses naturally fall into the "applied" and "management" categories, and then by the subject matter of test basics, cores, scan, built-in self-test, ATPG tool selection, and memory testing.

For the design or DFT engineer, the most important information is the items that describe how to establish a test methodology or implement particular DFT techniques, how the techniques affect the design budgets, and why they are needed (in many cases, the DFT engineer will need a course in "how to argue effectively"). These topics can be divided into the following courses of study: test basics, scan, AC scan, ATPG, memory test, built-in self-test, and core test.

For the manager, the most important information is the items that deal with trade-offs, tasks, schedules, and cost. These items also fall into similar categories of implementing scan or AC scan, adopting an ATPG methodology, implementing a BIST-based test methodology, and developing or using cores.

Test Fundamentals: If you wish to study just the basic fundamentals of testing, with the goal of learning about the test process, understanding the creation of a test program, applying the test program to a tester, and assessing the quality metric, then I would recommend reading all of Chapter 1, Sections 2.15.2, 2.16.1, 3.15.1, 5.4.1, 5.18.1, and 5.25.1, in addition to whatever training material is at hand for the target tester.

Scan Techniques: If you wish to develop a course of study in the understanding and application of scan techniques, then I would recommend reading all of Chapters 1 and 2 for a fundamental background into testing, faults, and ATPG, and then Sections 3.1 through 3.19 for a background in basic DC scan, and Sections 4.14 through 4.18 to explore the interaction of scan with memory architectures. If memory architecture needs to be learned to understand the "scan interaction with memory", then I also recommend reading Sections 4.1 through 4.5.

AC Scan Techniques: If you wish to extend the study of scan into the AC and at-speed scan realms, then I recommend the additional reading of Sections 3.19 through 3.26 to explore the differences between DC scan and adding the extra dimension of timing assessment.

ATPG: If you wish to understand automatic test pattern generation (ATPG) and perhaps wish to understand the development of an ATPG methodology, then I recommend reading all of Chapters 1 and 2 for a basic background of the test process and ATPG, and in addition, Sections 3.1 through 3.12 and Section 3.16, since understanding scan, partial scan, and scan design rules is fundamentally interrelated to the development and application of an ATPG methodology.

Memory Test: If you wish to develop a course of study in the understanding and application of memory testing, then I recommend reading Sections 4.1 through 4.13, and if the study should include a built-in self-test methodology, then 4.19 through 4.28 as well.

BIST: If you wish to develop a course of study in the theory, application, and use of built-in self-test (BIST), both logic and memory, then I would recommend reading Sections 3.3 through 3.11 for a scan background, Section 3.16 on scan rules, Section 3.22 through 3.25 on at-speed scan concerns, and 3.27 on logic BIST. In addition, I would recommend reading Sections 4.1 through 4.13 for a background on memory architecture and testing, and then read Sections 4.19 through 4.28 on memory BIST.

Embedded Core Test: If you wish to understand embedded core testing, either from the development of a testable reusable core, or from the integration standpoint, then I recommend reading all of Chapter 1 for a fundamental background in testing and test application; Sections 2.1 through 2.17 for a basic background in vector generation, vector optimization, and controllability and observability; Sections 3.1 through 3.26 to understand AC, DC, and at-speed scan and scan issues; Sections 4.19 through 4.28 to understand memory BIST and memory BIST issues; and the entirety of Chapter 5 on cores and core-based design.

Management: If you are concerned with the cost, schedules, tasks, and trade-offs of the various topics contained within this book, then a "management" course can be outlined that covers just the sections needed for quick understanding of the issues, costs, and so on. For example, a course of study in the trade-offs involved with DFT would include Sections 1.1 and 1.2 for an understanding of what is tested and why it is accomplished, 1.4 and 1.6 for an understanding of what is measured and with what type of testing, 1.8 to explore the cost trade-offs of test equipment, and 1.10.2 to understand the cost trade-offs involved with test programs.

To understand the selection and development of an ATPG methodology would require Sections 2.1 through 2.3 to understand what an ATPG methodology is and why it should be adopted, Section 2.14 to understand the difference between combinational and sequential ATPG, xxvii

Section 2.16 to understand ATPG vectors, Section 2.17 to understand the design rules that will be imposed on the design if ATPG is adopted, Section 2.18 on the selection of a tool, and Section 2.19 for a summary overview of all the issues.

To understand the adoption of a scan methodology would require reading the identified ATPG Sections (2.1-2.3, 2.14, 2.16-2.19) and scan Sections 3.1 through 3.3 to understand scan testing versus functional testing, Section 3.12 to understand full-scan versus partial scan, Section 3.16 to understand the design rules that will be imposed on the design for scan, Sections 3.19 and 3.20 to understand AC scan versus DC scan, Section 3.25 to understand the tasks involved with scan insertion, Section 3.26 to understand the selection of critical paths for AC scan, and Section 3.28 for a summary overview of all the issues.

To understand the adoption of a memory BIST methodology would require reading Section 4.1, which introduces memory testing; Sections 4.4 and 4.5, which address memory design and integration concerns; Section 4.6, which outlines the trade-offs between different memory test methods; Sections 4.19 and 4.20, which introduces memory BIST testing and requirements; Sections 4.22 through 4.24, which outline chip issues and concerns in applying BIST; and Section 4.28, which provides a summary overview of all of the issues.

To understand the issues, concerns, trade-offs, cost factors, goals, and tasks involved with developing a reuse core, or adopting an embedded core-based design style, I would recommend understanding the scan and memory BIST information, and then reading Chapter 5 in its entirety.

CHAPTER 1

Test and Design-for-Test Fundamentals

About This Chapter

This chapter is the Test and Design-for-Test (DFT) Fundamentals portion of the text and has been designed to teach the basic fundamentals of test and design-for-test. This material will establish the groundwork for understanding the test and DFT techniques and trade-offs described in the other chapters of this book.

Chapter I includes material on the definitions and terminology of test, the test process, design-for-test, fault models, test measurement criteria, types and styles of testing, test platforms, test programs, tester requirements, vectors, and vector processing.

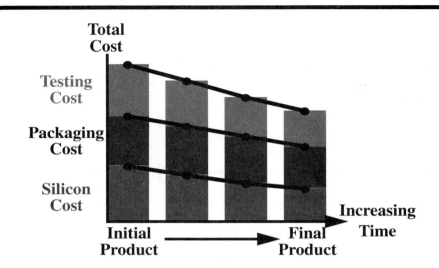

The goal over time is to reduce the cost of manufacturing the product by reducing the per-part recurring costs:

- reduction of silicon cost by increasing volume and yield, and by die size reduction (process shrinks or more efficient layout)

- reduction of packaging cost by increasing volume, shifting to lower cost packages if possible (e.g., from ceramic to plastic), or reduction in package pin count

- reduction in cost of test by:
    — reducing the vector data size
    — reducing the tester sequencing complexity
    — reducing the cost of the tester
    — reducing test time
    — simplifying the test program

Figure 1-1 Cost of Product

1 • Test and Design-for-Test Fundamentals

1.1 Introduction to Test and DFT Fundamentals

1.1.1 Purpose

This section describes testing and the test process, and introduces the concept of design-for-test (DFT), so that you will more easily understand the test and DFT techniques and trade-offs discussed in later sections in the book.

1.1.2 Introduction to Test, the Test Process, and Design-for-Test

The main purpose of the test process, as it is applied to the manufacturing of semiconductor products, is to provide a measure of the quality and/or reliability of a finished semiconductor product, whether that product is a stand-alone die or a packaged part. The purpose of design-for-test (*DFT*) is to place the "hardware hooks" on the die to enable the ability to conduct the quality-reliability measurement. If done correctly, the DFT will: enable the quality goals to be met with a high degree of confidence (high test coverage) during testing; allow the coverage measurement to be done efficiently and economically to meet the cost-of-test goals; and enable some form of test vector automation, such as automatic test pattern generation (*ATPG*) to help meet the aggressive time-to-market or time-to-volume manufacturing requirements.

Currently, the semiconductor industry is changing on many fronts:
- smaller geometric features on silicon;
- the reduction of internal voltage levels;
- the use of new physical processing techniques such as copper interconnect, low-K dielectrics, and silicon-on-insulator;
- the use of complex megacells and macrocells known as cores;
- the use, and reuse, of existing simple and complex "hard" macrocells;
- the integration of huge amounts of memory; and the integration of large amounts of logic on system-on-a-chip devices.

These industry changes make quality and reliability requirements more important than ever, but at the same time are creating aggressive challenges in the ability to measure the quality level, or to measure the quality level economically. For example:
- The smaller features, lower operating voltages, and new processing techniques are expected to create new classes of defects and failure effects, such that existing defect, fault, and failure models may be inadequate for detection and characterization.
- The ability to place large volumes of logic and memory on a single die in a short period of time is expected to increase the cost-of-test due to longer test times and larger amounts of test vector data, and is also expected to compress the design cycle time involved with the DFT portion of the chip.
- The ability to use complex cores and reuse macrocells is expected to create access and vector delivery problems with design elements that may be quite testable as stand-alone devices, but are to be embedded or doubly embedded within the overall chip architecture (macrocells or memory arrays are included within other macrocells that will be embedded).

In the past, the test process has been characterized as an "over-the-wall" event that occurred when the design team completed the design and threw it to a dedicated team of test and/or verification professionals. This test process was largely the translation and reuse of the functional simulation verification vectors to the target tester platform, and the manual creation of new verification vectors to get more coverage. The vectors were laboriously "graded" for fault coverage by conducting fault simulation. The "post design" time to provide a set of vectors to meet high quality expectations was measured in months and even years. The new test issues have now required that test structures be designed into the device to assist with the test process, quality measurement, and the vector generation. These changes are being driven by market pressures to: 1) provide high-quality parts (*Quality/Reliability*); 2) meet time-to-market or time-to-volume manufacturing windows (*TTM/TTV*); and 3) meet product cost goals by meeting cost-of-test goals (*COT*). Figure 1-1 illustrates this trend.

1.1.3 Concurrent Test Engineering

The rapid pace of process geometry reduction in recent times has led to a strange cost/pricing effect. Whereas in the past, the overall cost of the semiconductor product has been able to be cost managed to a lower level over time, lags in supporting technology have finally led to "smaller is more expensive." There is currently a reasonable die size limit to the ability to connect bonding wires from the package to the silicon. This has led to a somewhat "limited" die size for certain pin count packages. If a process geometry shrink is applied to an existing device, then the logic may not fill up the space within the pad ring. To eliminate this "white space," the chip design organizations are adding more functionality or memory arrays. These significantly increase test time and test vector data. If the memory supported on the chip, or the additional memory added, is a non-volatile memory (Flash, EEPROM), then test time is also significantly increased, since these types of memories require long test times naturally.

Another test cost impact has been the adoption of the "reuse" hard core macrocell (for example, an entire microprocessor as a megacell). For intellectual property (*IP*) security reasons, hard core macrocells from core providers are delivered complete with existing test vectors (having the ability to generate vectors against a structural description of a provider's core may give away all the proprietary internal features for the core). Semiconductor devices made with many hard cores require building test programs out of many vector sets delivered by the core providers. The vector sets may not be very efficient, or may not be able to be applied simultaneously to other vector sets. This fact leads to a high cost-of-test in having a complicated "patchwork" test program.

The only way to manage the ever-increasing cost-of-test problem is to apply design-for-test techniques to the device during the design phase—this concept has been called concurrent engineering in the past. In reality, the test needs of modern state-of-the-art devices will require that test considerations become an inseparable part of the design process, and the cost-of-test problem can be treated just like any other engineering budget (area, power, frequency, noise immunity, electrostatic discharge immunity, etc.) and techniques can be applied to manage it as part of the overall design (see Figure 1-2). The key optimization factors are the amount of vector data, the complexity of the tester operation and pin timing management, the expense impact of the target tester, the total "in-socket" test time, and the average quality level measurement per vector or per second (the vector efficiency).

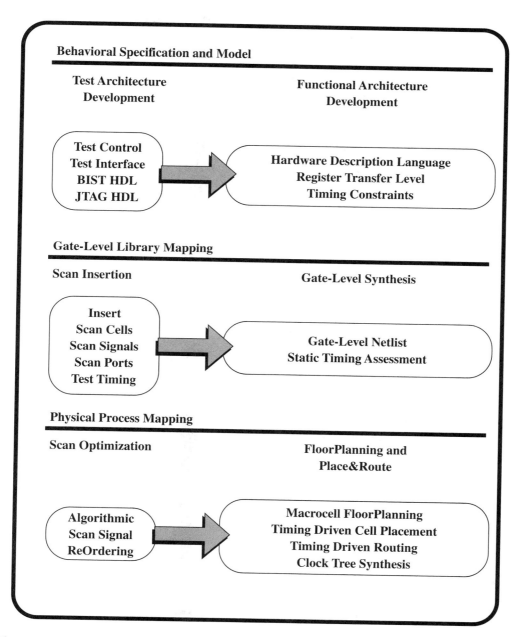

Figure 1-2   Concurrent Test Engineering

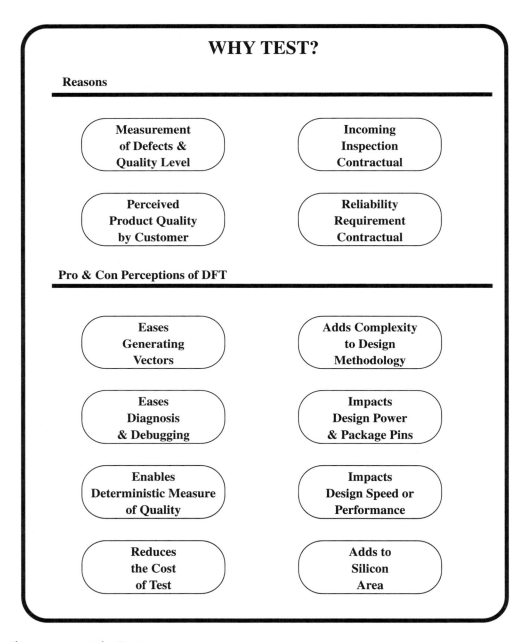

Figure 1-3   Why Test?

1.2 The Reasons for Testing

1.2.1 Why Test? Why Add Test Logic?

One of the most frequently asked questions that test and design-for-test engineers have to field is "Why do we have to add all of this test logic?—what is it for?—and why?" These questions may (and do) come from many different people and organizations involved in semiconductor device design—engineering, management, marketing, sales, manufacturing, and test. Design engineering is mostly concerned about the impact of extra logic for test on area and performance budgets; management wants to understand the trade-offs of the advantages of test features versus the schedule risk and implementation costs; marketing and sales want to understand the test feature set and what the "reportable" quality numbers are; and test engineering wants to know whether the test feature set is compatible with the currently supported test platform, and whether it meets or exceeds the "cost-of-test" budget.

Basically, there are two real reasons to conduct testing, as shown in Figure 1-3: it is contractually required by a specific customer (or by generic customers by way of an implied or actual warranty); or it is required by a competitive environment where the quality level and device reliability are among the items voted on by customer dollars.

Test logic may be added for several reasons. Some key reasons cited quite often are "to increase the test coverage" and "to reduce the time it takes to qualify the part" (deliver vectors that meet a certain test coverage). Many organizations are now beginning to understand that "reducing the cost-of-test" by supporting test features during the design phase is manageable, achievable, and can be treated as an engineering problem with a set of targetable goals. The act of adding logic or features to enhance the testability of a design is generally referred to as Design-for-Test (*DFT*). More specifically, DFT can be defined as adding logic or features to enhance the ability to achieve a high quality metric, to ease the ability to generate vectors, to reduce the time involved with vector generation, or to reduce the cost involved with the application of vectors.

1.2.2 Pro and Con Perceptions of DFT

Whenever testing and design-for-test is brought up as a subject, people and organizations have differing opinions and perceptions. For design engineering organizations and individuals, the perception of DFT is generally negative:

- it adds work and complication to the design methodology flow
- it negatively impacts chip design budgets such as
  - power
  - area
  - timing
  - package pin requirements
- it adds tasks and risk to the design schedule

However, for test professionals, the perceptions of DFT are usually positive and include such items as:

- having the ability to measure the quality level deterministically

- making it easier to generate the necessary vectors;
- making it possible to support all test environments easily
  - wafer probe
  - manufacturing test
  - burn-in
  - life-cycle
  - board-level integration
  - engineering debug
  - customer return debug
  - process characterization
  - yield enhancement and failure analysis
- allowing the cost-of-test to be reduced in all environments
  - reduces tester complexity (and cost)
  - reduces tester time (and cost)
  - reduces tester requirements (pins, memory depth, pin timing)

However, much of this perception is true only if the additional logic added for test is designed to meet these goals. In many organizations, some test logic is still added in an ad hoc manner to meet specific goals, which has limited payback in the larger scheme of things.

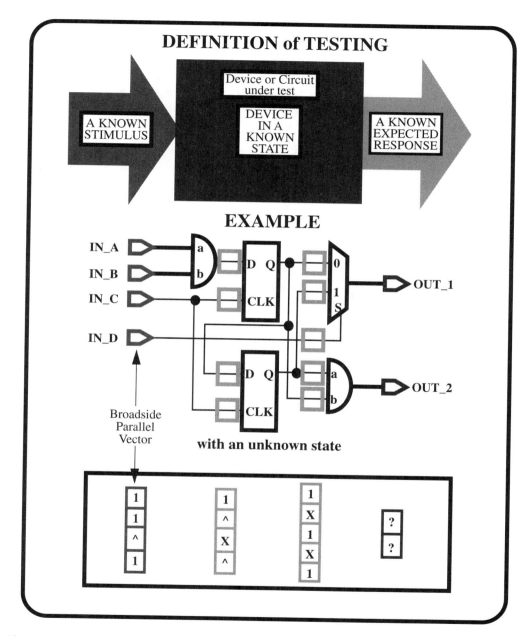

Figure 1-4   Definition of Testing

1.3 The Definition of Testing

1.3.1 What Is Testing?

The definition of testing is a rather simple concept and it applies equally to chips, boards, and systems: testing is possible when a *known input stimulus* is applied to a unit in a *known state*, and a predictable or *known response* can be evaluated. In general, as Figure 1-4 shows, the known response from a circuit is compared to an "ideal" or "golden" *expected response*.

1.3.1.1 A Known Input Stimulus

Testing can be accomplished on any level of circuitry from transistors to gates to macrocells to cores to chips to boards to systems. Any kind of testing requires putting a known input stimulus into the unit-to-be-tested's input pins. This input stimulus can be applied in a simulator, on a tester, or from other logic within the system. For digital systems, this stimulus is in the form of logic 1's and 0's known as vectors. Having the ability to apply the stimulus to the unit under test is generally referred to as *controllability*.

The key here is access to the inputs and applying a known value. If the unit-to-be-tested is buried within a chip or board, then getting the known stimulus values to the identified input pins may require modifying or translating vectors, to be applied at accessible locations through "tested-known-good" logic, or it may require special access logic such as test busses or test boundary rings.

1.3.1.2 A Known State

For testing to be valid, the unit-to-be-tested must be in a known state. It doesn't matter what stimulus is applied if the internal state of the circuitry is not known explicitly. The unit-to-be-tested must act as a simple linear operator—in other words, if "A" is applied, "B" happens (every time). This result is not true if randomness exists. The repeatability of operations is known as *determinism*, and the repeatable response of the unit-under-test is generally termed as being *deterministic*.

Placing a circuit into a known state can be done by applying a set of initialization values that will result in a known state unconditionally, or by applying some form of hardware set/reset that will leave the logic in a known state.

Randomness can legally exist during testing as long as the concept of the propagated "X" or the masking (ignoring) of outputs that are *indeterminant* is applied. The key point to be made here is that the action (or reaction) based on the inputs is deterministic on some subset of tested logic and that non-deterministic actions are either identified or ignored.

1.3.1.3 A Known Expected Response

The third requirement for valid testing is being able to evaluate a known expected response. There is no point in applying logic values to a circuit that has been initialized if the response cannot be evaluated or if it is not known whether the response is truly correct. For digital systems, this determination requires having a pre-conceived notion of the system function (e.g., a specification for the action, the truth table for a logic gate, etc.) or by simulation of vectors to get a set of expected response vectors against which to compare the circuit response. The ability to evaluate the output response of the unit-under-test is generally referred to as *observability*.

The key here is access to the output values and the ability to compare the circuit response to a known expected value. The same problems that affect applying a known input stimulus are valid here. The output response must be detected directly or translated through "tested-known-good" logic, or through a special logic access structure, to a point where it can be detected (observed), and then it must be compared to a pre-determined known value. If the response has randomness or non-determinism associated with it, then that portion must be ignored or masked.

1.3.2 Stimulus

The term *stimulus* or *input stimulus*, when referring to electronics testing, generally refers to one of two items: logical 1's and 0's, which translate to voltage levels for digital testing (for example, TTL levels refer to voltages of 0V minimum and 5V maximum to represent a logic 0 and logic 1, respectively); and analog waveforms that may be represented as functions of voltage or current. When applied as a grouping of some sort (for example, the set of 1's and 0's applied to multiple chip pins at one clock cycle edge), the stimulus is generally referred to as *vectors*. Usually, the term *patterns* represents collections of vectors bundled together into a single file.

1.3.3 Response

The *response* or output response of a device during digital testing may actually involve more that just a logic 1 or logic 0. In general, the concepts of unknown (indeterminate) and high impedance must also exist. These values are generally referred to as logic X's and Z's, respectively.

The majority of IC testing involves "voltage level" evaluation, but current measurement and time-based waveform analysis are also common descriptions of output response.

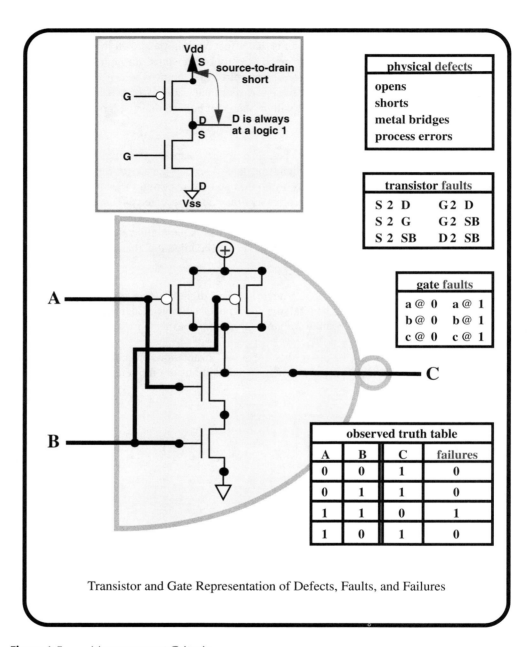

Figure 1-5   Measurement Criteria

1.4 Test Measurement Criteria

1.4.1 What Is Measured?

When a known stimulus has been applied to an initialized circuit and the response has been evaluated and is found to not match the expected response, then a failure has occurred. However, in the language of test, the terms *defects* and *faults* are used in addition to the term *failure*. Each of these terms has a different meaning in the test process, shown in Figure 1-5.

1.4.1.1 Defects

Defects are the physical problems that occur in silicon. Different processes and process geometries (feature sizes) have different groupings of problems that manifest themselves as defects. Some common silicon CMOS (complementary metaloxide semiconductor) defects are: gate-oxide shorts, insufficient doping, process or mask errors, metal trace opens, metal trace bridges (resistive and 0 ohm), open and plugged vias (anomalies in the vertical interconnections between metal layers), and shorts to Power (Vdd) or Ground (Vss).

1.4.1.2 Faults

A fault can be viewed as the failure-mode manifestation of the defect (faults will be covered in more detail in Section 1.5). In order to calculate a quality measurement, a measurement criteria or model has to exist. The fault is a model of the failure mode of the defect that relates the defect to circuit behavior. For example, a gate-oxide short effectively shorts the source of a transistor to that transistor's drain. This creates the transistor-level fault of a shorted "source-to-drain" (referred to in Figure 1-5 as S2D). These faults may be modeled by sticking the output of the transistor to a fixed logic value (logic 1 or 0). Other transistor fault models can represent the shorts or opens in the "gate-to-drain" (G2D) or in the "source-to-substrate" (S2SB) portions of the transistor. The behavior resulting from these models may be high current, a high impedance state, or intermittent behavior.

If the transistor is a required element used to resolve the logic value on the output of a gate, then the output of the gate may stick to a single logic value (logic 1 or 0). The model or failure mode of the example NAND-gate shown may be to relate the defect to the transistor failure mode (fault model) if the level of analysis is "switch-level." However, if the transistor-level is not the level used for fault measurement, then the failure behavior must be related to the gate-level failure mode (fault model). In the case of the gate-level description, the model most used for quality measurement is the single-stuck-at fault.

Another example of a defect mapping to a fault is the small metal open on a connection trace. In this case, the logic value can still propagate, but by a tunneling current across the open. The propagation of the logic value is slowed as if it were passed through a series resistance. The fault model at the circuit level is to represent the defect as a unit of signal delay along the length of the signal trace. This fault model is generally known as a delay fault.

1.4.1.3 Failures

Many defects may exist, and these defects may all map to fault models, but a defect, or a transistor-level fault, or a gate-level fault may not represent a failure. The term *failure* has two requirements that must be met: 1) that the fault model be exercisable and observable; and 2) that failure measurement criteria be established.

For example, if a gate oxide short on a transistor (S2D = Source-to-Drain short) within a NAND-gate results in the output of the NAND-gate always being a logic 1, then the fault model is a gate output stuck-at 1 (C@1 in Figure 1-5). For this gate oxide defect, represented as a transistor-level fault model, which is then represented as a gate-level stuck-at fault model, to be classified as a failure, the NAND-gate must be exercisable and observable (and exercised in such a way that the transistor defect is activated), and the fail condition must be established.

In this case, if gate input A=1 and gate input B=1 can be applied to the NAND-gate (to force the output to logic 0), and if the fact that the output did not resolve to the logic 0 state can be observed through the other system logic, then this fault can be classified as a failure. However, if the circuit is constructed with redundancy so that the NAND-gate cannot be exercised to the proper state or so that the NAND-gate response cannot be observed, then the fault exists, but not the failure. The fault may be classified as undetected, blocked, tied, redundant, or untestable, and may or may not be counted in the fault detection metric. Some examples of faults that are not failures would be:

- small delay faults on paths with large timing margins
- stuck-at faults on redundant or reconvergent fanout logic
- current leakage less than an established level

For these examples, the chip would have a defect content, but not a fail indication. The manifestation of the defects may be merely a reliability problem, which should be screened out as infant mortality during an early exercise process such as burn-in (accelerated life-cycle testing by continuous operation of the device in an environmental chamber or oven).

1.4.2 Fault Metric Mathematics

A common metric used as quality assessment on digital circuits is *fault coverage*. Fault coverage is related to another term known as *vector grading* (also more commonly referred to as *fault grading*). The measurement of fault coverage is done with reference to a set of vectors. The vectors are said to exercise and detect some measured number of faults out of all possible faults in the design.

The general mathematical equation for fault coverage is:

(Total Detected Faults) / (Total Fault Population)

For example, (9821 detected faults)/(10,000 total faults) = 98.21% fault coverage.

There is a warning about fault coverage, however. Different tools may report different fault coverage numbers. Sometimes the absolute total number of faults in the design is not used as the denominator of the measurement process—in these cases, the denominator may be reduced by the number of "untestable faults" due to the circuit configuration (redundant, tied, blocked, constrained, etc.), since this content is constant and based on the design. The discrepancy in fault coverage also occurs when different tools identify faults differently so the total fault population changes. For example, some tools may classify only stuck-at 1's and 0's on gate inputs and outputs at faults, whereas other tools may classify stuck-at 1's and 0's on gate inputs, gate outputs, and the interconnecting wire nets as faults.

1 • Test and Design-for-Test Fundamentals

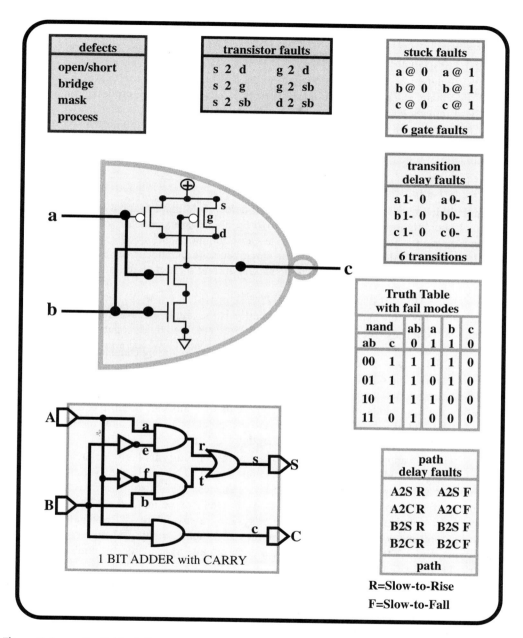

Figure 1-6 Fault Modeling

1.5 Fault Modeling

1.5.1 Physical Defects

Physical defects are what really happens to the part in the physics realm. At the chip level, these fall into many categories: interconnect defects, packaging defects, process problems, and so on. These defects have their source in such things as local contamination, lattice shifting, incorrect masks, process errors, insufficient doping, silicon die stress, and interconnect errors or dicing stress. The task of a failure analysis engineer to isolate manufacturing defects and determine their source so that yield may be maintained at an acceptable level.

1.5.2 Fault Modeling

Fault modeling is the translation of physical defects to a mathematical construct that can be operated upon algorithmically and understood by a software simulator for the purposes of providing a metric for quality measurement. The most common fault models supported in modern VLSI and core-based digital design are the single-stuck-at DC model, the transition and path delay AC models, and the toggle and pseudo-stuck-at current measurement models.

1.5.2.1 The Single-Stuck-At Fault Model

Currently, a popular model for digital circuitry is the single-stuck-at fault. This is a DC (static) approximation whereby all failures are represented as an individual gate-level pin port or wire net connection that acts as if it were shorted to Vdd or Vss. However, a limiting assumption is applied to the single-stuck-at fault model known as the "single-fault assumption." This assumption requires that only one fault exist at a time during any analysis, assessment, or diagnostic. The assumption is applied to reduce the complexity of the assessment (the "what if" of extrapolating multiple faults in designs with thousands of faults can become overly compute-intensive to the point of impossibility with current computer technology). This model is applied regardless of frequency or time domain considerations.

Even with this approximation simulating a circuit to identify its faulty behavior is a compute-intensive process. Historically, as IC sizes approached VLSI levels, the switch-level (transistor) simulation of the past gave way to gate-level. This change is also true of the fault model. It used to be sufficient to fault model at the transistor level (for example, the S2D—"source-to-drain" short). Now with circuits moving into the "system-on-silicon" realm, even gate-level fault simulation is too time-consuming. It would be far too difficult, costly, and time-consuming to translate exact models of all of the physical defect sources into a simulator that could verify analog electrical and physical behavior.

In the move from "switch-level" fault assessment to gate-level fault assessment, many corporate organizations justified the move to what was viewed as a "less accurate" fault coverage assessment, by using the derivation of the gate stuck-at fault from the transistor-level fault. If a connection to one of the transistors used within a gate is open or shorted, then the result is that the output of the gate is a constant logic 1 or logic 0 (for example, the gate-level output C stuck-at-Vss—C@0 fault model in Figure 1-6). In a few cases, a transistor fault may force the gate output to a high-impedance (Z) or float value, and switch-level simulators can handle this effect (but for the most part, this behavior is not a common fault model).

The key to the gate-level stuck-at fault model is that a fault will cause a boolean gate error which, if observed, is classified as a failure. For example, for the NAND-gate in Figure 1-6, if either input A or B is stuck at logic 0 (AB@0), then the output C will always produce a logic 1 output; if input A is stuck at a logic 1 (A@1), then the output C will produce an incorrect output—a logic 0 instead of 1 when the A=0, B=1 input combination is applied; if the input B is stuck at a logic 1 (B@1), then the output C will produce an incorrect output—a logic 0 instead of a 1 when the A=1, B=0 input combination is applied; and finally, if the output C is stuck at logic 0 or logic 1, then the output of the NAND-gate will always be a logic 0 or logic 1, respectively, regardless of the applied input to the gate.

1.5.2.2 The Delay Fault Models

There are two generally supported fault models for AC effects (that is, fault models that include the concept of timing)—the transition delay and path delay fault models.

The Transition Delay Fault Model: The basic timing fault model is the *transition delay* model, which can be viewed as a modified version of the single-stuck-at DC model, but with an additional restriction applied to allow it to be used to assess the time domain. The extra step is to force the gate-output to the expected fail value at some time period prior to the observation event, and to then apply a transition and to conduct the observe or sample event at a defined period of time. The requirement of being observed at the fail value at an initial time and then being observed at the pass value at a second time maps a time assessment onto the simple DC stuck-at model. Instead of viewing any node, net, or gate input as a stuck-at logic 1 or 0, the transition delay fault model applies the slow-to-rise (the logic 0 to logic 1 transition) or the slow-to-fall (the logic 1 to logic 0 transition).

The application of the fault model to gate pin ports or wire net connections is identical to the stuck-at DC gate-level model, except that the stuck-at-0 and stuck-at-1 values are now termed zero-to-one and one-to-zero transitions (for example, the gate output C zero-to-one transition—C0-1 in the transition delay table in Figure 1-6). The transition delay fault model is also known as the *gate delay fault model,* since the observation of a fault can be attributed to a slow gate input or slow gate output (the delay can be related directly to a gate in the modeling sense).

The Path Delay Fault Model: The other AC fault model in common use is the *path delay model*. This model can be viewed as the sum (or stack up) of combinational gate transition delays along an identified circuit path. The overall model is similar to the transition delay model, but instead of a single gate pin port or wire net connection being targeted, an entire path made of multiple gate pin ports and wire net connections is the target of the fault model. Sometimes some confusion occurs about what a "path" fault is. Some fault assessment tools view the collection of individual faults making a circuit path as the target of the analysis, and some tools have a path naming convention that implies that the "named path" is the fault. In reality, different vector generation and fault simulation tools may have different accounting methods in dealing with the processing of faults—the key here is that transitions or "vector pairs" can be applied to (or launched into) individual gates, or whole described circuit paths.

The restriction of having a fail value being observed at the output of the path at some initial time, and then having the pass value being observed at the path output at a defined later time, is still required. The path transitions are generally referred to as a concatenation of nets from the input transition point to the output observe point. For example, the path A-to-S slow-to-rise (A2SR in the path delay table in Figure 1-6) may be formally described as pin and net elements, "A-a-r-s-S Rising" (similarly, A2SF means path A-to-S slow-to-fall).

1.5.2.3 The Current-Based Fault Models

Another popular set of fault models that has gained acceptance is the current-based fault models. Some defects have been found to map to fault effects that are not caught by either the stuck-at or the delay fault models. These defects may cause high current leakage (as may the defects that are represented by the stuck-at and the delay fault model). Some organizations argue that a defect that does not cause a boolean error is not a fault or a failure condition, but is more of an indicator of a reliability problem (a condition that does not fail now, but may indicate an early failure due to a mechanism such as metal migration or an active short that will eventually create an open). The classification of a high current leakage condition as a failure condition is subjective to each organization's quality requirements, but in some cases with critical applications (such as pacemakers), any out of tolerance behavior must be classified as a failure.

The Pseudo-Stuck-At Fault Model: This current-based fault model is also largely based on the DC stuck-at fault (a net, node, or gate connection may be represented as a short-to-Vdd or a short-to-Vss). What makes the pseudo-stuck-at model different from the stuck-at model is the observation event. The stuck-at model is based on observing the voltage value representing the boolean logic values 1 and 0, and in conducting the observation of the logic value of the fault effect at a voltage observe point (a test point, a scan flip-flop, or an output pin). This is not the goal for the pseudo-stuck-at model—the goal is to only drive the fault effect to the output of a gate and to then conduct the observation event by conducting a current measurement of the supply or return current being used by the device (the current delivered to a device under test may be measured from the supply side, or it may be measured from the return side of the power supply).

In CMOS logic, quiescent current measurement is effective because non-switching gates can be modeled as only using quiescent or diode reverse current—the leakage current used by a static non-switching device should be very low (the sum of the reverse currents for the number of transistors on the device), any active bridging, shorts, and some open defects should result in a current measurement that is an order-of-magnitude greater than the "quiescent current rating."

The Toggle Fault Model: Another fault model used for current measurement is the toggle model. The toggle model is based on being able to place every net, node, and gate connection to a logic 1 or a logic 0—this is a very powerful simulation tool used to verify that every component of a design description can be driven to both logic values. The toggle model was a natural to be used for the current measurement technique since the goal is to exercise as many bridges or shorts as possible. This model is one of the easiest to apply, since it attempts only to place nets, nodes, and gate connections to a logic value—there is no attempt to drive values to an observe point.

Quiet Vector Selection: However, an extra restriction is applied to fault models used for the current measurement technique. This requirement is that the design must be in a static or quiet state when the observation event occurs. This means that the design must not have any active current sources and sinks (such as pullup and pulldown logic) that can compromise (overwhelm) the detection of the fault. Since the current measurement is based on the "quiescent" current draw, which can be a very low limit such as 10 microAmps, then any dynamic or active logic can easily overwhelm this limit by drawing milliAmps of current.

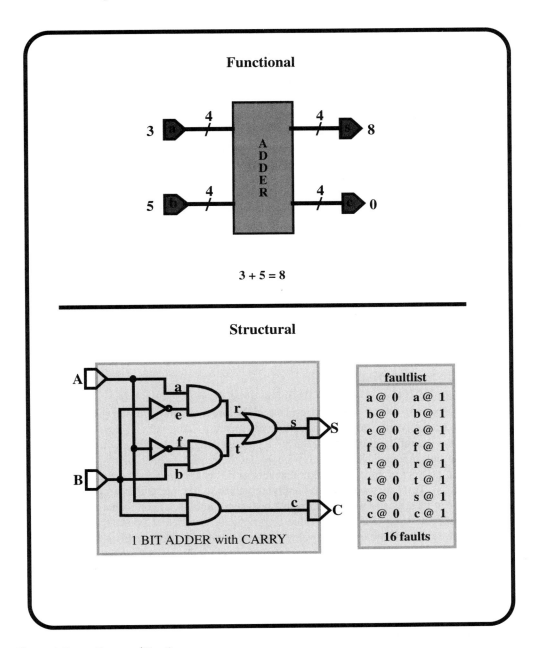

Figure 1-7 Types of Testing

1.6 Types of Testing

1.6.1 Functional Testing

A functional (or behavioral) test is used to verify that the model or logic behaves as it was intended (specification behavior). This may also be called a "design verification test." If the function represents an adder, then tests will be written to see whether the logic commits the necessary add functions and nothing more (see Figure 1-7). For example, if 3+5 and 4+4 are the only values specified or critical to the adder, then these are the values applied to test the adder's functionality—if all values are important, then all combinations will be applied. Functional testing is measured by the logic committing the correct action (known expected response) to the (known) applied stimulus—100% functional correctness is the standard expectation and this should be verified at the behavioral (HDL/RTL) level, or the gate-level of the design with a simulation process.

Functional testing may also include timing or power consumption as part of the functional standard. In the case of timing, a second dimension is added to the behavior of the circuit: 1) that it conduct the correct behavior; and 2) that it conduct this behavior to a timing standard. For example, if a general purpose adder is placed within a chip design and it must be able to add any two hexadecimal numbers within one clock cycle, then the goal is to provide an additional set of tests that verifies that the slowest operation conducted by the adder is still faster than the target cycle time of the device. In the case of power consumption, some defined operations must occur and the device can consume no more power than the specified power budget.

1.6.2 Structural Testing

A structural test is used to verify the topology of the manufactured chip (see Figure 1-7). Another way of saying this is, "Given a good circuit before the manufacturing process, structural testing can be used to verify that all connections are intact, and that all gate-level truth tables are correct after the manufacturing process." This type of testing can be done with reliance on the static stuck-at fault model. This is an assumption that the physical defect being searched for will represent itself as a net or gate connection that acts as if it is always a 1 (short to Vdd) or always a 0 (short to Vss). Tests are developed by applying values to the inputs that toggle the suspected defective node to its opposite value (forcing a 1 on a stuck-at 0 node) and then applying values at the inputs that would allow the good value to propagate to a detect point. If the value at the detect point differs from the expected value for a good circuit, then a fault has been detected. Structural testing is measured by fault coverage (how many faults were detected compared to the total number of possible faults)—95% to 99.9% is the standard expectation in the semiconductor industry in general. A delay fault model can be applied similarly to assess timing structure, and a current-based fault model can be used to assess power consumption.

1.6.3 Combinational Exhaustive and Pseudo-Exhaustive Testing

Combinational exhaustive or 2^n testing is used to verify how the combinational portion of the model or logic behaves when every possible set of values is applied to the "n" input ports, even if some of the vectors applied have no functional significance. Pseudo-exhaustive testing is the application of some portion of all possible 2^n logic values (a shorter test). Exhaustive and pseudo-exhaustive testing is done when a piece of logic is being characterized (timing or logic operation); when one type of model is being compared to another (such as a gate-level equivalent circuit and its transistor model); or when vectors will be applied in some manner without an understanding of the circuit (for example, a Built-In Self-Test methodology where random seeming logic values may be applied externally to the device, or where logic is embedded within the chip to create random or pseudo-random logic values). Exhaustive, pseudo-exhaustive, and full-exhaustive (see next subsection) are measured by the logic producing the correct logical or boolean response to the applied stimulus and comparing the response to a known ideal (golden standard) or expected response—100% functional correctness and timing within a window of parameters is the standard expectation.

1.6.4 Full Exhaustive Testing

Full exhaustive testing or $2^{(n+m)}$ testing is the same as combinational exhaustive testing except that there are sequential elements that hold state embedded within the model or circuit. Simply applying every possible combinational value to the input pins is not enough to characterize a sequential design. There is the added complication of the applied sequence of all possible values of stimulus. A state machine with M elements requires 2^m tests to test all sequences. To fully test a combinational circuit with 2^n applied values, and to also consider all possible sequences, the combinational 2^n input values must be multiplied by 2^m state sequences, resulting in $[(2^n)*(2^m) = 2^{(m+n)}]$ tests. Again, this type of testing may be applied for characterization purposes, or when vectors are applied without an understanding of the circuit.

1.6.5 Test Styles

Each of the described types of testing is usually related to a "style" of testing. For example, functional testing is accomplished by applying "broadside" or "parallel" behavioral or operational vectors to the device, whereas structural testing is mostly accomplished using a test style or structured test technique known as "serial scan," and exhaustive testing is usually accomplished using another structured test technique known as built-in self-test (BIST).

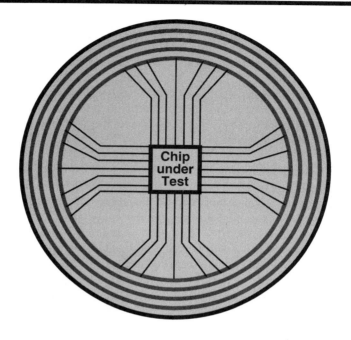

The chip will be accessed by the tester at its pins only

A custom (load) board will be made for this purpose

Each pin has a limited number of bits available (e.g., 2 MB)

The test program (set of vectors and tester control) will be applied at tester speed (may be less than actual chip speed)

The primary goal of manufacturing test is structural verification

Figure 1-8   Manufacturing Test Load Board

1.7 Manufacturing Test

1.7.1 The Manufacturing Test Process

The manufacturing test is accomplished using Automatic Test Equipment (ATE). A big mistake made by many design organizations is to view the tester as a hardware version of the simulator. The tester is not a simulator! The tester has limitations in application of vectors and clocks, in measurement of response, in application and measurement of frequency and timing, in support of the number of channels or pins, and in support of vector memory depth. These make the manufacturing test process a non-ideal process (in the mathematical sense, as compared to a simulator).

1.7.2 Manufacturing Test Load Board

The manufacturing test is applied to the chip through a load board (see Figure 1-8). The outer edge and underside of the load board contains the contacts that interface with the tester, and board interconnect routing connects this edge to a socket contained in the middle of the board. Each connection of the load board to the tester has a given amount of tester memory available. For example, common ATE sizes are 1, 2, and 4 MB (millions of bits per pin).

Any special features needed to test the chip can be placed on this board. An example would be to multiplex two separate tester memory channels to one tester input pin to make twice as much vector memory available, or to apply data at twice the frequency (switching data from one channel in the first half of the clock cycle, and from the other channel during the second half of the clock cycle).

The load board must also reflect the performance of the device it is intended to test. If it is a high-frequency design, then board routing must observe the rules of transmission lines for length, width; and impedance matching. The load board should not set the frequency limit of testing, the tester's internal limitation should apply.

The key point to be made here is that the tester only has access to the chip by its pins and the load board may be the limiting factor for the level of test that can be applied.

1.7.3 Manufacturing Test Program

The manufacturing test program is a piece of software that contains the stimulus vectors, response vector, masking information, and what is needed to control and sequence the tester. This vector data and sequence information usually comes from the post simulation log or an ATPG tool and the test engineer who structures or builds the program.

In order for the fault coverage reported by the fault simulator or ATPG tool to be accurate, the fault assessment must emulate the tester as best as possible. In other words, the stimulus points used for the fault simulation should be the chip's input pins, the detect points should be the chip's output pins, and the application timing and detect strobe timing should match that of the tester. This matching will insure that the fault assessment (e.g., fault simulation) reports the same fault coverage that the tester is actually capable of detecting.

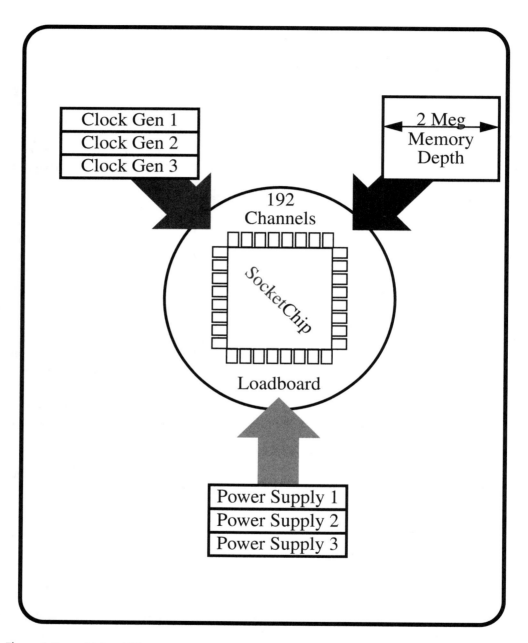

Figure 1-9 Using ATE

1.8 Using Automatic Test Equipment

1.8.1 Automatic Test Equipment

Integrated Circuit Automatic Test Equipment, also known as IC ATE, is generally a very expensive (i.e., greater than $1,000,000.00) test platform. The ATE can be modeled or represented as a number of channels with memory depth (a channel basically representing the memory behind a package pin), a number of clock generators, and a number of power supplies (see Figure 1-9). These resources are applied to the chip through a loadboard that houses the chip socket. Some other resources or functions may exist, such as:

- memory test functions (MTF's) to conduct algorithmic memory tests
- analog-to-digital (ADC) and digital-to-analog (DAC) converters
- current measurement devices for Idd and Iddq testing
- frequency counters to verify or synchronize output clock signals

The cost drivers of ATE are mostly related to the number of channels, the amount of memory depth, the frequency of operation, and the accuracy of measurement.

1.8.2 ATE Limitations

ATE is not exactly the same as a simulator. A simulator can operate with or without timing, can be cycle-based or event driven, and can access any signal, variable, or constant in a design. The waveforms applied by a simulator can be viewed as the ideal mathematical representation of a signal, unknown signals can be represented as a logical construct known as the "X," and the high impedance state can be represented as the logical construct known as the "Z."

The ATE, on the other hand, can operate only at the edge of the design through the package pins, and has real signal and clock delays and signal degradations, chip output loads, load board transmission line and round trip signal timing, and thermal considerations. The tester also has signal edge placement precision, accuracy, and edge-rate (rise and fall time) limitations.

1.8.3 ATE Cost Considerations

The main component of the "cost-of-test" is the time a chip spends "in-socket" on the ATE. There is a minimum cost based on the minimum time that it takes an automated handler to insert and remove a chip from the tester socket (production throughput). Aggressive handler time is generally in the 1-3 second range. Cost reductions to bring tester time below the handler threshold require testing multiple die in parallel with the same tester.

If the test programs applied on the tester are excessively complex or contain an excessive amount of vector data, then a vector reload may be required. A reload adds a large amount of time to the overall test process, which increases the "in-socket" time, and therefore, the cost of test. The ways to minimize the test program complexity and data size are to apply simplifying techniques, such as single-edge-set testing, and to apply vector compression to the vectors sets. A simpler test program also reduces the complexity of the ATE that is required. A less complex ATE requirement means that a less expensive tester can be used.

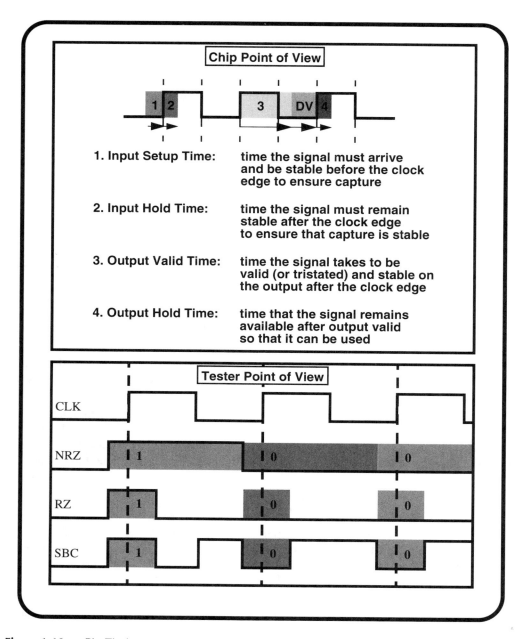

Figure 1-10   Pin Timing

1.9 Test and Pin Timing

1.9.1 Tester and Device Pin Timing

The tester must apply input stimulus to the input pins of the IC package, and must measure or strobe the output response, with a known relationship to a reference. The most general form of reference is the clock signal(s) created by the tester and applied to the device-under-test as the system clock(s). The application of input stimulus and the measurement of response data with respect to the reference is generally known as pin timing (see Figure 1-10).

Sometimes the device under test has an internal clock generator (e.g., PLL), and the tester must align the application of input information and the measurement of output information against a clock-out signal from the chip itself. Since the timing relationship between the package pins, the internal logic, and the tester can be very subtle, and the accuracy and precision and limitations of the tester must be factored into the analysis, sometimes a tester must conduct an extensive edge search and update a variable timing table to ensure that applications and measurements with respect to the clock are accomplished within the stated specifications. This type of operation consumes test time and can be very costly as compared to referencing timing specifications from a tester generated clock (a "clock in" reference).

1.9.2 Tester Edge Sets

The cost-of-test is lower if the test process requires the support of fewer edge sets (the maximum optimization occurs if all testing can be done with only one edge set). The timing point, with respect to the reference, at which new information is applied to a device-under-test input pin, or at which response data is observed on a device-under-test output pin, establishes a timing edge set. If a package pin supports multiple application or strobe times or if the package pins support application or strobe time based on multiple references (e.g., clocks), then that pin is said to require more than one edge set. Another way of saying this is if the tester must support more than one timing standard or waveform to conduct testing, then this is another edge set.

An edge set is not just the pin timing standard, but also may include the tester timing format. The tester timing format has to do with the way the tester generates or manages the signal. Single edge set testing generally means that data is applied, or measured, on a pin with the same timing standard and the same tester format throughout the test program.

1.9.2.1 Device Pin Timing Requirements

The device-under-test must have the vectors applied from the tester based on the chip's own timing limitations or specifications related to its input pins, output pins, and clock reference signals. These specifications are described in the chip's User Manual or Specification Data Sheet and represent such items as: the timing point associated with the application time that a stable signal must be applied to input pins with respect to the upcoming clock edge to ensure that it will be correctly captured by sequential elements within the part; or the timing point at which data exiting output pins will be stable and valid with respect to the previous clock edge. The pin specifications are:

Input Setup: the minimum time point before a reference clock edge that input data must be applied and stable to ensure that the pin data will be captured by the chip.

input Hold: the minimum time point after a reference clock edge that stable input data must remain to ensure that the input data captured on the reference edge does not change.

Output Valid: the minimum time after a reference clock edge that exiting output data will be stable and valid for use by the system outside the chip.

Output Hold: the minimum time after the valid time point, or from the next reference after the edge that the valid time point is based on, that output data will remain valid and stable for use by the system outside the chip (how long will the data remain valid).

Tristate Valid: the minimum time after a reference clock that an output pin will assume a stable high impedance state.

1.9.2.2 Tester Pin Timing Formats

Tester pin timing formats (see Figure 1-10) are a way for testers to easily generate digital signals and represent them in the test program. These formats also inherently represent the concepts of "print on change," or "print on time." For example, one way to represent pin data is to represent it as a "Non-Return" format—this format emulates the print-on-change format in that the pin data changes value only when the data really changes, even if it is several clock cycles before the data changes. As a contrast, a "Return" format will always return the pin data value back to some default value at the end of the reference signal interval (representing the print-on-time format). Some common pin data formats are:

Return-to-Zero (RZ): The drive signal will always be a logic 0 unless the tester is asked to drive a logic 1—the logic 1 data will be applied and will be valid between the pin's defined input setup time and the pin's defined input hold time.

Return-to-One (R1): The drive signal will always be a logic 1 unless the tester is asked to drive a logic 0—the logic 0 data will be applied and will be valid between the pin's defined input setup time and the pin's defined input hold time.

Return-to-Complement (RC): The drive signal for data applied between the input setup time and the input hold time will be defined by the vector data, and the tester format will always return to the complement of the data after the hold time.

Non-Return-to-Zero (NRZ), *Non-Return-to-One (NR1)*, *Do-Not-Return (DNR)*: The non-return formats are basically "print-on-change" data formats that keep the same applied value throughout the period, and will change the data only when the vector data changes state, even if this is several clock periods. The data changes occur at the pin's defined input setup time.

Surround-by-Complement (SBC): The drive signal for data applied between the input setup time and the input hold time will be defined by the vector data, and the tester format will always surround the data—before and after—with the complement of the data.

1.9.3 Tester Precision and Accuracy

The tester itself provides a limitation to validating the timing of the chip-under-test. This is the tester's ability to resolve timing measurements or applications, and the ability to place the edges used as data and reference.

Precision: The smallest measurement that the tester can resolve.

Accuracy: The smallest ambiguity in edge placement.

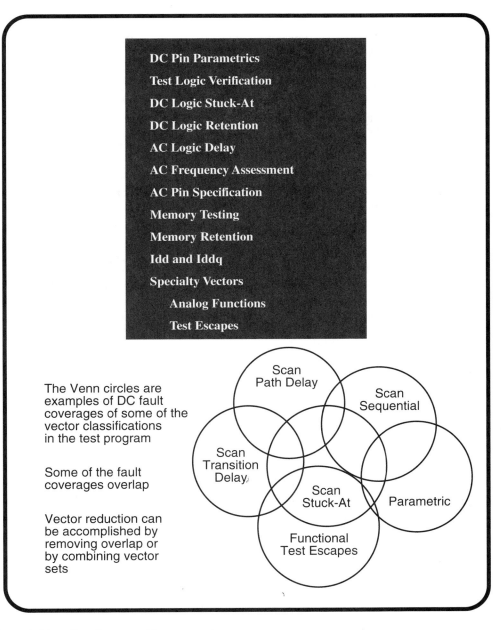

Figure 1-11 Test Program Components

1.10 Manufacturing Test Program Components

1.10.1 The Pieces and Parts of a Test Program

The test program contains more than just the "functional verification vectors" that are delivered by the design team. Most test programs, regardless of the device under test, contain the same type of testing functions (vector products) or test patterns groupings. Note that the purpose of a manufacturing test program should be for structural verification of the silicon, and not for "functional design verification" (or as I am fond of saying, "If you haven't proven it's a microprocessor before you spend all that money for a mask set, then it's your folly"). The test program should be "economically optimized" to provide a cost-effective and efficient method to verify that the manufacturing process did not add errors or defect content to the finalized design. The test program should not be used as a vehicle to verify the final design or design library assumptions (unless the chip under test is the process technology vehicle—the first design to be made in a new process, or with a new design library).

The following general vector products are contained in a common test program (see Figure 1-11) and should be arranged to provide an early exit if the test program finds a failure—this is known as stop-on-first-fail and can significantly increase tester throughput if the test program is long compared to the handler index time: DC pin parametrics, DC logic stuck-at, DC logic retention, AC frequency assessment, AC logic delay, AC pin specification, memory testing, memory retention, Idd (power), Iddq (leakage), and specialty vectors.

1.10.1.1 DC Pin Parametrics

The very first tests applied to the chip are usually the DC pin parametrics (also referred to as "shorts and opens," "DC Levels," and "Pin Leakage" testing). The purpose of parametric testing is to verify:

- the connectivity of the package pin bonds; the power and ground structure of a chip
- the pin voltage levels
- the pin current leakage

In some cases, the test can be used to measure the soundness (leakiness) of the silicon. Parametric testing is accomplished by conducting DC measurements of the voltage and current specifications at the chip interface. Some of the measured specifications are:

Iil, Iih: the measurement of input current leakage with the pin driven to the low and high (0 and 1 logic) voltage values.

Iol, Ioh: the output current leakage measured when the pin is driven to the low and high voltage values. Note: if a global tristate signal is used, then these measurements may be replaced with an *Iz* where the pin is at the high impedance level.

Vil, Vih: the measurement of input DC voltage levels associated with logic trip points.

Vol, Voh: the output DC voltage levels associated with logic trip points.

1.10.1.2 Test Logic Verification

The test logic needs to be verified for structure and operation before this logic can be used to accomplished testing.

1.10.1.3 DC Logic Stuck-At

The majority of chip logic is general chip control logic. This logic is tested by the application of vectors, and the vectors are graded against the stuck-at fault model. Traditionally, these vectors have been functional operation vectors or scan vectors that are run at slow frequencies compared to the chip's maximum rated frequency.

1.10.1.4 DC Logic Retention

If the chip is declared to be a static design, or if there is a concern that sequential logic elements will undergo capacitive discharge or some other form of voltage degradation, then a logic retention test involving the stopping or slowing down of the reference clock may be accomplished.

1.10.1.5 AC Frequency Assessment

If the chip is declared to operate at some frequency range or at some max frequency (for example, in the chip's data sheet), then this declaration must be verified. This verification has traditionally been done with the application of functional operation vectors that are similar to the vectors to be used in system operation. However, in the past, no general tool has been available to apply a real grade to these vectors that would assess the probability that every timing-limiting path has been exercised.

The more modern method is to generate scan vectors that apply the functional-cycle at-speed (the shift in of the scan state may be done at a slower frequency). These vectors are generated from paths extracted from static timing analysis, and are limited to the timing measurement of register-to-register transfers. The vectors associated with all paths that violate or just meet timing slack are all that are needed.

1.10.1.6 AC Logic Delay

As the process geometry shrinks, delay defects become more prevalent and dominate the fault content. Functional or scan vector sets applied at-speed need to be used to detect this fault content. A number of vectors that would exercise and detect a wide distribution of delay fault content need to be applied.

1.10.1.7 AC Pin Specification

Similarly, the pin specifications also limit the frequency, but these measurements are associated with the setup, hold, data-valid, and tristate-valid measurements applied to the input and output pins. Traditionally, these have been functional vectors, but scan vectors are beginning to be used here as well.

1.10.1.8 Memory Testing

Any embedded chip memory arrays must also be tested. However, memory testing is different from general logic testing. Memory testing is defect-based and uses algorithmic vector generation to exercise the many defect classes. The test applied may be a functional pattern to get an on-board controller to test the memory; it may be a direct memory test function (MTF) if there is a direct memory access architecture; or it may be a clock-only vector if an on-chip BIST is evident.

1.10.1.9 Memory Retention

Memory arrays are regular structures that are constructed to be more dense than the general logic structures. To this end, the possibility of leakage is more probable, so the loss of data has a higher likelihood. Retention testing involves placing data into the memory arrays and either stopping the clock for a period of time, or locking the memory during test and reading the value back at some later time in the test process.

1.10.1.10 Idd and Iddq

Most digital testing is based on voltage measurements. A different point of view involves using current measurements. One method is to measure the supply (or return) current while the chip is operating dynamically (Idd Dynamic—chip power measurement), and another method is to apply a state and stop the clock and measure the leakage or quiescent current (Iddq or Static Idd).

1.10.1.11 Specialty Vectors

Not all digital designs are wholly digital, and not all digital logic is testable. If specialty logics, such as analog functions, are contained within the chip, then they must be tested as well. Similarly, if logic on the chip is not testable by the normal test methodology, then different techniques need to be used to get this test coverage.

Analog Functions: non-digital logic such as Digital-to-Analog convertors (DAC), Analog-to-Digital convertors (ADC), voltage controlled oscillators (VCO), and phased locked loops (PLL) also require test and verification. The testing conducted here may require non-digital test methods such as voltage, current, and timing waveform measurements, frequency counting, clock dithering, and hysteresis measurements.

Test Escapes: In some cases, logic is left untested by the basic test methodology. For example, the memory arrays may be tested by BIST, and the general combinational and sequential logic may be tested by scan. The scan logic may test only up to the level of registration before the memories, and the BIST may test from the memory array boundary inward—this restriction may leave some logic between the scan and memory array not tested. In this case, a functional vector may have to be delivered that will exercise the paths between the general logic and the memory arrays, since neither methodology covers this logic. Unfortunately, the term *test escape* implies that the chip was delivered faulty and that the problem was discovered by the end user, and this result is sometimes true—a vector is then crafted and added to the manufacturing test program to ensure that new chips do not have this identified problem.

1.10.2 Test Program Optimization

A test program may be composed of several pattern/vector sets. If the amount of test vector data exceeds the tester memory or test time budget (or the cost-of-test budget), then some test program optimizations may be required. Two basic optimizations can be applied: 1) test data optimization; and 2) test program control or sequencing optimization.

1.10.2.1 Vector Overlap and Compression—Test Data Optimization

Each vector set has a fault coverage metric. Since the manufacturing test program should be a structural device (used to test the structure of a device), then pattern sets with overlapping structural test coverage can be considered redundant. These vector sets can be optimized by removing vectors with redundant coverage, or by combining the portions of tests that are not redundant.

The usual methods of vector data reduction are winnowing, compression, and compaction. Winnowing is the removal of redundant pattern sets. Compression is usually thought of as a form of winnowing, but within a pattern set—the reduction of the size of a pattern set by removing redundant vectors. Compaction is the combining of multiple vectors into single vectors by mapping the "X" or "don't care" states of the different vectors so that the new vectors contain more data but have no logic conflicts.

1.10.2.2 Single Edge-Set Optimization—Test Sequence Optimization

Each tester has an operating language that allows different pattern sets to be included and applied with different clock frequencies, voltages, and pin timing. A significant amount of test time involved with a test program is the amount of control and sequencing time required by the tester. Another cost-of-test optimization is to restrict all testing to a single timing edge set so that the test program sequencing overhead is minimized (all vectors seem to look alike).

1.11 Recommended Reading

To learn more about the topics in this chapter, the following reading is recommended:

Abromovici, Miron, Melvin A. Breuer, Arthur D. Friedman. *Digital Systems Testing and Testable Design.* New York: Computer Science Press, 1990.

Dillinger, Thomas E. *VLSI Engineering.* Englewood Cliffs, NJ: Prentice Hall, 1988.

Gulati, Ravi K. and Charles F. Hawkins, eds. I_{DDQ} *Testing of VLSI Circuits—A Special Issue of Journal of Electronic Testing: Theory and Applications.* Norwell, MA: Kluwer Academic Press, 1995.

IEEE Standard Tests Access Port and Boundary-Scan Architecture. IEEE Standard 1149.1 1990. New York: IEEE Standards Board, 1990.

Parker, Kenneth P. *The Boundary-Scan Handbook, 2nd Edition, Analog and Digital.* Norwell, MA: Kluwer Academic Publishers, 1998

Tsui, Frank F. *LSI/VLSI Testability Design.* New York: McGraw-Hill, 1987.

CHAPTER 2

Automatic Test Pattern Generation Fundamentals

About This Chapter

This chapter contains the Vector Generation Fundamentals and is included so that the limitations and rules to be placed on logic during the design-for-test (DFT) development stage can be understood. The information in this chapter is also presented so that the measurables and trade-offs involved in the selection process for vector generation automation tools can be understood.

This chapter is not a comprehensive exhaustive text on the mathematical theory behind Automatic Test Pattern Generation (ATPG), nor is it an industry review of the current state of the art in available ATPG tools. This chapter is a basic treatise on the fundamental actions of ATPG tools and the related analyses based on industry use and support of currently available tools. This chapter includes material on the ATPG flow and process, fault models, fault analysis, fault equivalence, fault masking, fault simulation, stuck-at and delay fault-based ATPG, current-based ATPG (iddq), multiple time frame ATPG (sequential and combinational), vector compression and compaction, and the trade-offs (benchmarking) involved with the selection of vector generation tools.

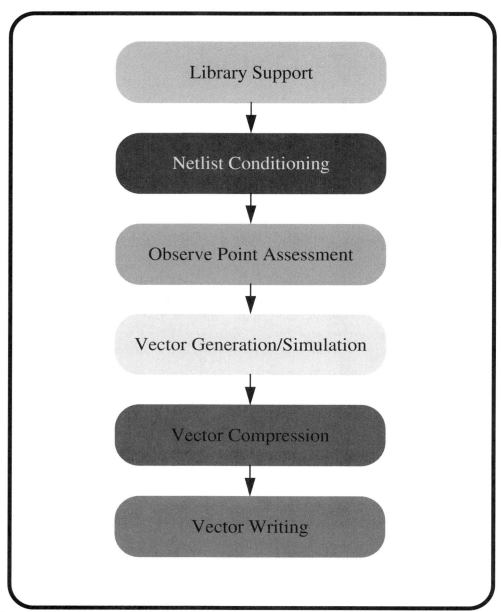

Figure 2-1 The Overall Pattern Generation Process

2.1 Introduction to Automatic Test Pattern Generation

2.1.1 Purpose

The Automatic Test Pattern Generation (ATPG) chapter describes the basic fundamentals of the tool-based algorithmic test pattern generation process so that you will more easily understand the algorithms, abilities, and limitations of the process in the context of the rules and restrictions placed on the hardware design. This chapter will also provide the fundamental basis for understanding the test techniques and trade-offs involved with the hardware test architectures discussed in later chapters and sections of the book. In addition, this chapter will also describe the key "measurables" involved with selecting and using an automated vector generation process and ATPG tools so that the correct tool and supporting tools may be chosen to meet the cost-of-test, time-to-market, and quality level goals of a semiconductor provider.

2.1.2 Introduction to Automated Test Pattern Generation

The cycle time associated with the design process is continually shrinking. An aggressive time-to-market or time-to-volume manufacturing requirement (TTM/TTV) means that a heavy reliance on design and test automation needs to be supported. A high quality requirement means that a more extensive analysis of the circuit must be done to achieve a high test coverage result. A low cost-of-test budget means coverage-efficient and memory-compressed vectors (vectors that will fit in the available target tester's memory so that a vector re-load is not required). These goals can be met by establishing an overall automated test pattern generation methodology based on using a software tool for vector or pattern generation.

In the semiconductor design and test support marketplace, many different tools are in use for vector generation. Some of the tools are proprietary and internal-only to a company and some are publicly available from CAD/CAE vendors. There are also differences in the tools in their abilities and levels of support. Some tools are combinational-only, and so they must be used on combinational circuits or on circuits that have been rendered effectively combinational-only (such as full-scan circuits), and some tools are fully or partially sequential, and so these tools can be used on sequential circuits with or without added DFT. ATPG tools also differ in the type of fault models they support. The more common fault models are the stuck-at, transition delay, and path delay fault models for logic (voltage-based) analysis, and pseudo-stuck-at and toggle for leakage (current-based) analysis. Also note that the different tools may also have different environmental and operational requirements: the type of operating system supported; the amount of RAM or disk usage needed; the type or size of CPU required; and the amount of time required to finish the job (runtime).

Establishing an automated vector generation process is not just in the selection and application of an automatic test pattern generation (ATPG) tool. A number of other support processes must also be provided or developed in addition to an ATPG tool. For example, the design description (behavioral model or gate-level netlist) must be of a format that the ATPG tool can use, and the vectors that are provided by the tool must be in a format that can be converted to the tester's data format.

2.1.3 The Vector Generation Process Flow

A complete and comprehensive automated vector generation process is made of several sub-processes, not just the ATPG tool (see Figure 2-1). A good ATPG tool, however, may include several of these sub-processes as related features. The sub-processes can be broken down into three basic categories:

- Pre-ATPG
- ATPG
- Post-ATPG

2.1.3.1 Pre-ATPG

The pre-ATPG section includes the tasks of creating the ATPG tool's library of standard cells, conditioning the design description so that the ATPG tool can operate on the provided data format, and establishing the goals and constraints of the test process.

As an example, if an ATPG tool will operate on the EDIF (or Verilog, or VHDL, or proprietary format) description of a gate-level netlist, then a library must be made that describes the gates used in the EDIF format (e.g., flip-flops, AND-gates, OR-gates, multiplexors, etc.) in terms of the ATPG tool primitives. If constructs in the EDIF netlist can't be modeled or supported with the ATPG tool library, then the EDIF netlist itself must be modified (for example, areas comprised of raw transistors, or of analog logic must somehow be represented by gate-level devices). All representations, in the library or in the design description, must have identical boolean behavior or else the created vectors may mismatch (i.e., provide erroneous results) when simulated against other design formats, or when applied against the silicon device.

Finally, before an ATPG tool can be used, the "test process" must be mapped onto the ATPG tool so that the vectors created will match the operation of the test platform. This step requires understanding the specific restrictions placed on the design by the tester, and by the design's own test architecture. In most cases, this is a "sequence" file that describes the sequence and pin values required to place the design in test mode, and describes which signals or pins must have certain logic values applied during the sequence. This type of file is generally referred to as a "procedure" or a "constraint" file. Also required is a file containing the description of the legal observe points so that the test coverage metric is based on the observations made by the tester and not by "virtual test points" in a simulator.

Be sure to note that some ATPG tools have extensive DFT or ATPG design rule checks (DRC) to determine whether the tool can operate on the design description. The application of DRCs and the "debug" and "repair" of the design description are generally considered part of the ATPG process, not the pre-ATPG process.

2.1.3.2 ATPG

The ATPG process is the actual operation of the tool against the design description to generate vectors. This process may occur several times during the design process as a prototyping step to determine budget compliance (e.g., number of vectors, fault coverage), or the process may be applied several times against the final design description to generate the various individual vector sets needed to provide all the pattern sets for the various different test modes and constraints. For example, a set of vectors may be generated with an on-chip bus being in the "read" mode, and another set of vectors may be generated with the bus being in the "write" mode (if the placing of the design into the read and write modes is too complicated for the ATPG tool to understand, or if the changing of the mode is not a stable process—for example, the mode change may require removing and re-applying power to the device, a power-up reset, that destroys or randomizes the current state of the device).

Do note that ATPG tools are not the "end all" solution for test coverage. The tools are limited to their algorithms, and some of the algorithms are not "strong" enough for some design problems. For example, combinational logic vector generation algorithms may not be designed for interaction with memory logic or extensive register files. So the tool may not have a way to place a specific value in a memory location, and to retrieve it later during an observe or control event also orchestrated by the ATPG tool. The limitations associated with each different ATPG tool are generally described as the required "design rule set."

2.1.3.3 Post-ATPG

After the ATPG tool produces the vector set, then you have the problem of having efficient and cost-effective vectors and mapping them to the target tester. In some cases, the ability to compress the vectors into a more compact form, so that they more easily fit in a tester's memory space, may be contained within the ATPG tool (an additional optimizing algorithm may be applied during or after the vector generation process). However, sometimes an ATPG tool will provide only simple vectors, and all optimizations must be applied as a post process.

The most common process applied to generated vectors is to convert them from a simulation data format to a tester format. Most of this type of software is generated internal to each company, but some "third party" (CAD/CAE provided) translation software does exist.

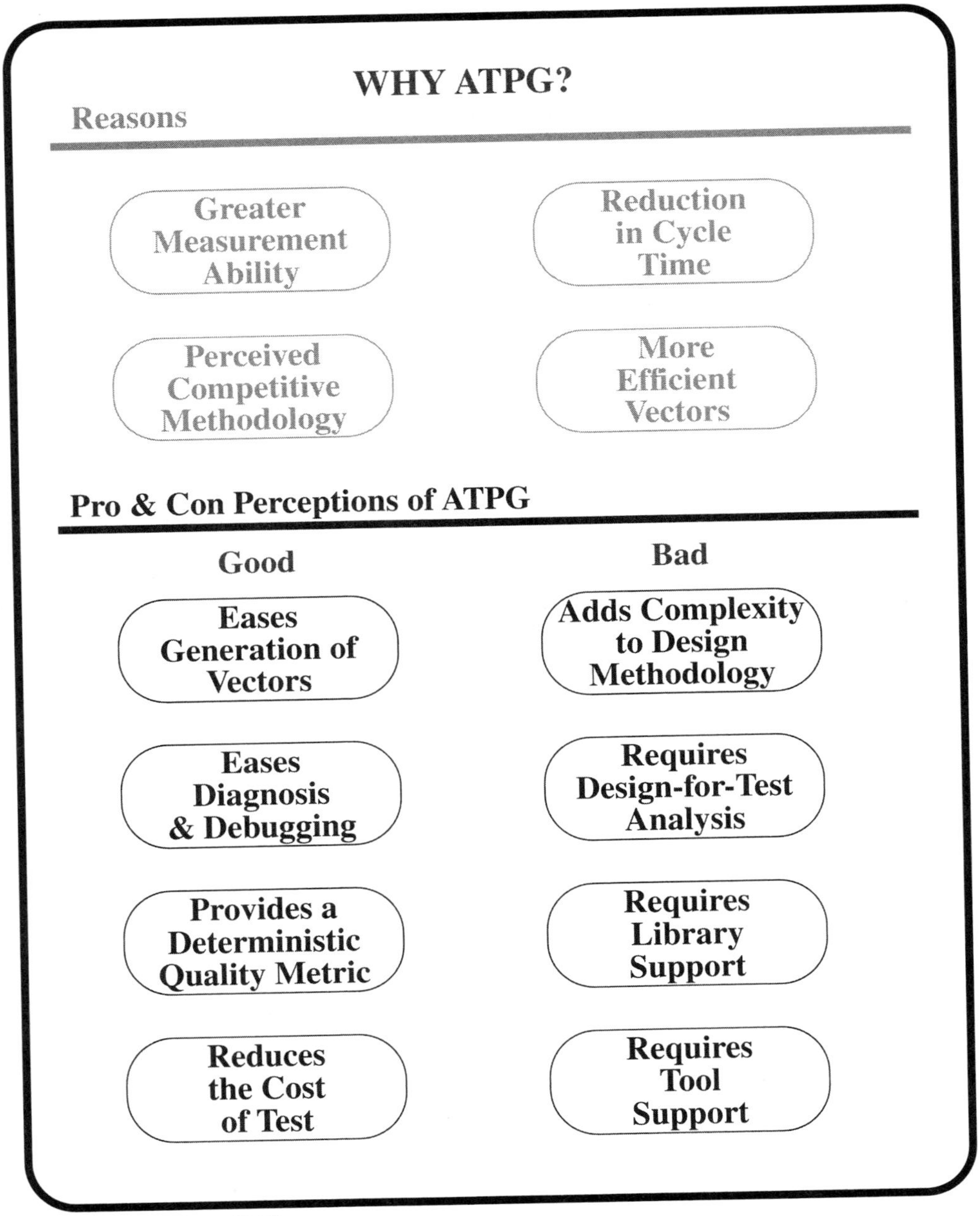

WHY ATPG?

Reasons

- Greater Measurement Ability
- Reduction in Cycle Time
- Perceived Competitive Methodology
- More Efficient Vectors

Pro & Con Perceptions of ATPG

Good
- Eases Generation of Vectors
- Eases Diagnosis & Debugging
- Provides a Deterministic Quality Metric
- Reduces the Cost of Test

Bad
- Adds Complexity to Design Methodology
- Requires Design-for-Test Analysis
- Requires Library Support
- Requires Tool Support

Figure 2-2 Why ATPG?

2.2 The Reasons for ATPG

2.2.1 Why ATPG?

Each year, the size and complexity of new semiconductor designs grow (a well known adage known as Moore's Law states that microprocessors will double in performance every 18 months). The semiconductor growth factors are in the number of transistors on a die, the maximum clock frequency range for device operation, the number of pins supported by a package, and the complexity of logic structures contained within a die (e.g., memories, analog conversion, bus interfaces, processing engines, clock generators). The business side of the industry also has growth factors, as shown in Figure 2-2: high quality and reliability requirements, reduced design-cycle time and time-to-market requirements, and reduced per-part cost/pricing. The business factors have been driving aggressive concepts such as rapid prototyping, reusable design, core-based design, systems-on-a-chip, and embedded operating systems, all of which result in limited test access (the amount of internal logic or transistors visible per package pin).

Historically, design and test engineers all have scary stories about the 8 to 18 months required to hand-generate the functional vectors to achieve some required test or fault coverage level needed to qualify a chip (achieve a quality level that allows the chip to be transferred to the customer). These engineers have also developed an intimate relationship with the drudgery task known as fault grading (simulating the wealth of functional vectors against their design description with a fault simulator). Those engineers that have used an ATPG process have discovered that the automated process is more of a "days to weeks" application that is applied by someone who knows the tool, not someone who is intimately familiar with the design (and I will make a disclaimer right here—I do personally know of some designs that have taken months and months with the ATPG process because of the lack of adherence to the required test design rules).

The other reasons to support the automated vector generation flow are in the delivered vectors. Traditionally, the ATPG vectors are much more efficient than hand-generated (or machine-generated) functional vectors, because they are "structure" based and provide the most amount of fault coverage in an optimized amount of clock cycles (ATPG vectors for a full-scan design, for instance, have a much higher fault coverage per clock cycle rating).

2.2.2 Pro and Con Perceptions of ATPG

In an analysis done similarly to that applied to the adding of test logic (DFT), the use and adoption of an ATPG process can also be found to have some positive and negative perceptions outlined in Figure 2-2. On the positive side, automatic test pattern generation is an automated process, so it reduces the time spent generating vectors to achieve a deterministic and measurable quality level. This reduces the overall cycle time involved with getting a part to the customer. Since the vectors have been generated against known fault models, then the vector failing on the tester can be traced directly to the faults involved with the generation of the vector. This can ease debug and diagnostics involved with determining the cause of the vector's failure on the tester (if the DFT design rules were followed). ATPG vectors have also been shown to be more efficient (fault coverage per clock cycle). This efficiency means fewer clock cycles on the tester, which can translate to a smaller test program and a faster test program—both of these reduce the cost of test.

However, some detractors from the ATPG approach view the adoption of an automated methodology in a negative light. The arguments here are that the adoption of a tool and the associated test methodology complicate the design methodology (in the front end) and may negatively affect the design schedule. The support of a vector generation tool and other associated DFT tools requires tool and library support. The negative aspects here are that support requires extra headcount (hiring more employees) or a loss of direct schedule control because of reliance on software vendors. The argument used most is that some form of structured DFT or adherence to "test imposed" design rules is required (which limits the field of solutions that the design team can explore and apply). Most of these arguments are based in the fact that the DFT and ATPG processes are now part of the design flow; they are no longer an "over the wall" test effort where the problems are transferred to a test organization after the design phase is over.

In reality, the adoption of an automated vector generation methodology depends on the ultimate goal of the chip design organization. If the product is leading-edge, extremely high-frequency, or pushing the bounds of size, then a large amount of custom logic and many manual tasks will be involved in the design of the chip. If the time to market window is not overly compressed and the cost-of-test is not a major concern (the high selling price of the part leaves plenty of margin for the cost-of-test), then the application of the old "tried and true" functional vector generation may be a viable solution.

However, if the chip is to be designed using an ASIC-type design flow based on behavioral models and synthesis and place&route tools, or if cost-of-test (a part with a lower price and lot less margin), time-to-market, and time-to-volume are major concerns, then an automated approach to vector generation is required.

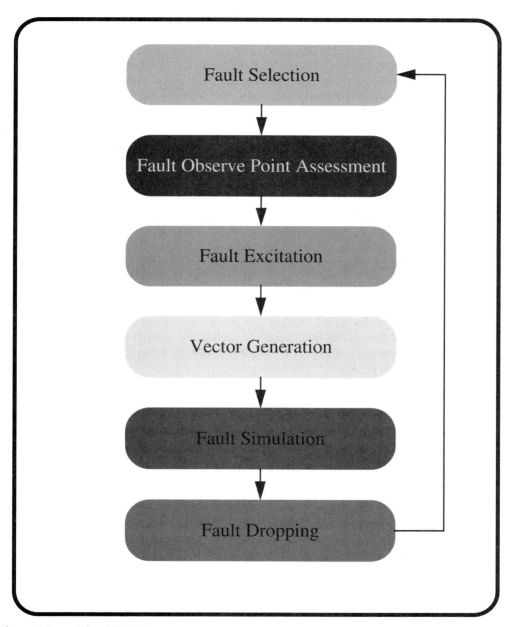

Figure 2-3 The ATPG Process

2.3 The Automatic Test Pattern Generation Process

2.3.1 Introduction to ATPG

Automatic Test Pattern Generation (*ATPG*) is the application of an algorithmic-based software to generate vectors and vector sets. The term *ATPG* is generally applied to fault-based or structural vector generation (vectors generated based on fault detection criteria); however, it has been applied (or misapplied) to some functional vector generation (such as creating vectors for the IEEE 1149.1 [JTAG] standard against a behavioral model description [BSDL] of the boundary scan architecture). This section will describe the ATPG process with respect to structural test vector generation only.

2.3.1.1 The Central ATPG Process

As was mentioned earlier, the overall ATPG process includes some tasks that must be done before and after the application of the ATPG tool. These are tasks that include preparing the computer environment for the tool, preparing the tool to accept the design description, preparing the design description for analysis, and processing the vectors created and written out by the tool to meet budgetary goals and to align with the data and control format of the tester. Right in the middle of all this is the "tool generates vectors against the design description" part of the process. This is the part of the process that most organizations and designers identify as the ATPG process, shown in Figure 2-3.

The central part of the ATPG process also has a sequence or flow. This flow has been developed over time with applied optimizations to continually improve the "fault coverage," the "runtime," and the "simplicity of use" of the process. So the fault-based ATPG process consists of many steps to provide an optimized methodology.

The first step after processing the design description involves establishing the fault model to be used, and the faults to be tested must be enumerated (listed by some criteria). Then a first fault to be operated on must be submitted to the ATPG software engine for vector generation.

The cone of logic that the fault is contained within must be assessed, evaluated, or analyzed to find a propagation path for the fault effect to an observe point. The step of finding an observe point is generally one of the more difficult analyses, since it involves forward tracing through the logic from the fault location to a location where the fault effect may be observed by a test platform. Next, the cone of logic is again analyzed, however, this time from the fault location backwards through the design description (netlist) to controllability points, in order to establish the logic value needed to excite the fault. The process of vector generation also includes backtracing from the elements on the observe path to other control points in order to enable the observe path to allow the fault effect to be propagated (all this tracing will be described later in sections 2.10.2 and 2.10.3). The order and the nature of the analyses for the propagation paths, excitation paths, and the path justifications are the differences in different applied algorithms.

The vector data is completed when the apply data, the expect data, and the mask data are placed in a data format (for example, an ASCII or binary format, or an industry standard format such as Verilog, WGL, STIL, etc.). The vector, however, may not be valid until it is simulated and verified. The simulation associated with vector generation is a fault simulation that determines whether the chosen fault is detected, identifies illegal conditions (such as driven contention—multiple drivers enabled and placing different logic values on a single wire net), and also identifies all other faults that are detected by the newly created vector. If the vector is valid, it is stored in the vector database, and any extra detected faults are dropped from the fault list to help reduce the overall amount of work the ATPG engine must accomplish. However, if the vector contains "real" or "perceived" contention, then the vector may be discarded.

The process then continues by selecting the next non-dropped fault in the list, and will continue until all the faults in the list have been processed.

2.3.1.2 Fault Models Supported

An optimal, efficient, and cost-effective ATPG process will allow for automatic pattern generation to meet the needs of much of the manufacturing test program (and in some cases, the vector generation can support the other test environments as well). This means that ATPG should allow for both AC (timing-based) and DC (timing-independent) voltage-based testing using the Combinational Stuck-At, Transition Delay, and Path Delay fault models. A good ATPG methodology should also support a current-based testing methodology by using either the toggle or pseudo-stuck-at fault models, or the quiet-vector-selection criteria to conduct Idd (power) and Iddq (leakage) tests.

Note: although HDL-based (behavioral) automatic vector generation for compliance against industry standards is available, this will not be discussed in this text as ATPG. Only structural vector generation will be classified as ATPG in this unit. An example of HDL-based vector generation would be an IEEE 1149.1 (JTAG) behavioral vector set for compliance verification.

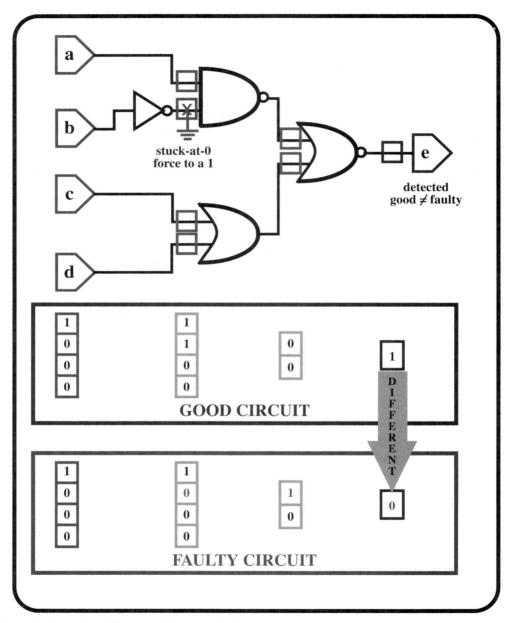

Figure 2-4 Combinational Stuck-At Fault

2.4 Introducing the Combinational Stuck-At Fault

2.4.1 Combinational Stuck-At Faults

Several types of fault models have been developed to analyze the design description, and to ultimately verify the silicon for defect content. Early fault models began by describing the failure possibilities of the transistor. When the design descriptions grew too large for a transistor analysis to be "schedule or runtime economical," then gate-level fault models were adopted. One of the first successful gate-level models was the stuck-at fault. When this fault model is applied to combinational-only logic (no sequential logic is part of the analysis), then the fault model is called the combinational stuck-at fault model (see Figure 2-4). The combinational-only analysis is a simplification for ATPG engines, since adding sequential behavior to the design description complicates the vector generation algorithms and the amount of data that must be considered during the analysis. The combinational-only type of fault model is generally applied to logic circuits that are constructed with a structured testability architecture such as full-scan.

When a structural test is used to verify the topology, and not the timing, of the manufactured chip, the fault model used is the stuck-at fault model, because the stuck-at fault model verifies that each net, node, and gate connection is not stuck at either a logic 1 or a logic 0. This type of analysis ends up verifying the connectivity and truth table description of each gate element.

In combinational logic, the stuck-at fault model represents itself as a gate pin connection, or connecting net value that is at a constant applied value of CMOS voltage levels of Vdd (logic 1) or Vss (logic 0).

2.4.2 Combinational Stuck-At Fault Detection

The existence of a defect mapping to a combinational stuck-at fault on silicon, in reality, may or may not immediately affect the logic it is associated with, and the fault effect may or may not translate through the circuit to modify the expected output. The combinational stuck-at type of fault is activated when the faulty pin, gate connection, or node, during functional or test operation, assumes a logic value that is the opposite value from its described stuck-at faulted value—for example, when a net described as "stuck-at logic 0" must propagate a logic 1 value (no information is learned when the stuck-at 0 net must propagate a logic 0). The combinational stuck-at fault is detected when the circuit behavior at the observe point is different from the expected value and this difference is due to the propagation of the fault effect (and not some other reason). What this means is that the gate function may be erroneous (faulty), but the fault is not detected unless the fault effect is seen at a circuit observe point.

A combinational stuck-at fault that is detected by a vector that is designed to "activate" or "exercise" the fault and then propagate the fault effect to an observe point is said to have been "deterministically detected." A combinational stuck-at fault that is detected on silicon is a failure.

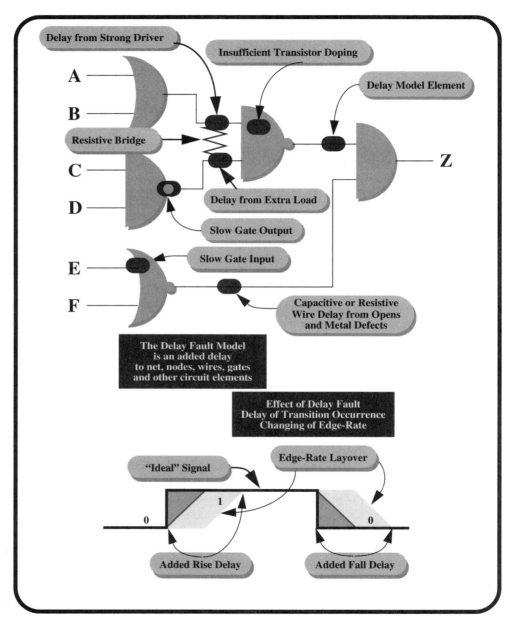

Figure 2-5 The Delay Fault

2.5 Introducing the Delay Fault

2.5.1 Delay Faults

In a real silicon design, the fault universe can't be described wholly by a static fault model that does not include timing effects. Therefore, a fault model was invented that was similar to the stuck-at fault, but included timing. So when a structural test is used to characterize the timing of the manufactured chip, the fault model used is the "delay fault" model (see Figure 2-5). In combinational logic, the single "gate" delay fault represents itself as a pin value of a gate element that acts as if it has a slow-to-rise or slow-to-fall logic transition, or as if an interconnect signal has a greater than normal propagation delay associated with it. There is also a "path" delay fault which is similar to the gate-delay fault, except that a whole path through several gates and net connections resolves the slow-to-rise or slow-to-fall concept to the last gate in the path (it can also be described as the subtle "stack up" of individual gate delays).

Many sources and types of delay defects at the silicon level can be modeled as delay faults. Some of them are:

- resistive gate-oxide shorts (a source-to-drain transistor short)
- open wire traces that support tunneling current
- open and plugged vias
- insufficient transistor doping (weak drive strength)
- wire traces with bridging (0 ohm or resistive voltage contention)
- extra metal capacitive loading on wire traces

2.5.2 Delay Fault Detection

The timing propagation delay that can be characterized as a delay fault can take two main forms: an added propagation delay where the logic transition happens later than expected, or an edge-rate degradation where the rise or fall time takes longer than expected (and some combination of both is also valid). The delay can be added at one location in the circuit (gate delay), or it can be the sum of many subtle delays that stack up to form a lumped delay at the observe point (path delay).

This delay can be evaluated and characterized only as a chip failure if the design description includes a a concept of time from some beginning point in time and a location in the design description, to some ending point in time and a different location in the design description, and if the delay can modify the behavior to be observed at the ending point. Generally, the fault model is described as the addition of some quantity of delay to a fixed (known) delay of a gate, or some element along a path, or to a data transfer along a path from one sequential device to another. The good circuit will propagate the correct and expected logic value to the observe point within a time window—the faulty circuit will slow the propagation of the expected logic value so that it will not reach the observe point in the given time window. The idea is to convert a timing delay into a boolean difference (error). Generally, the launch of the test value is synchronized from the rising edge of one clock cycle to the rising edge of the next clock cycle, or the rising edge of a clock in a different clock domain. These types of clock-to-clock transfers relate the propagation delay to an event the tester can understand, a clock edge (a clock cycle-to-clock cycle event or a clock domain-to-different clock domain event).

For example, if a register-to-register data transfer is done along some described path in a circuit and both registers are on the same clock signal, then the fixed timing window that a propagation event must meet is one applied clock cycle (with the data propagation occurring from the clock-to-Q of the first register to meeting the input setup time of the second register). The test is started by clocking a new and different value into the first register. If the addition of delay faults to the described path would cause the overall propagation delay of good data to exceed a clock cycle, then the fail condition would remain in the second register (thereby converting a timing fault into a boolean error).

As another example, in an asynchronous transfer from the beginning of a combinational cloud of logic to the end of a combinational cloud of logic, there may be a total time delay. The measurement of a delay fault would be with respect to the natural propagation time expected—if the asynchronous transfer exceeded some fixed time, then the fault would be detected.

Note that the "failure" condition is time-dependent and not all delay faults are failures. A small delay defect can add a small amount of delay, and the defect may not add enough delay for the circuit to fail the "timing measurement." For example, if a register-to-register transfer is a 10ns (100MHz) event and the real propagation time (sum of the delays of the gates and nets) of the circuit is 6ns, then a delay fault of up to 4ns may exist that will not fail the test condition.

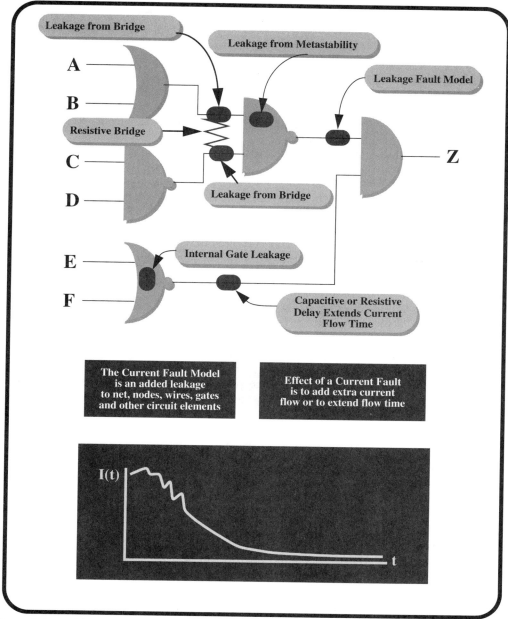

Figure 2-6 The Current Fault

2.6 Introducing the Current-Based Fault

2.6.1 Current-Based Testing

Both the stuck-at fault and the delay fault are based on propagating logic values to an observe point. The reliance on measuring a "logic" value bases the testing, and the fault model, in the "voltage" measurement realm. "Logic value" is just another mathematical model for a "voltage level." However, when a silicon product is being tested, voltage is not the only "measurable" applied. The tester's power supply also provides an electrical current flow. This leads to the concept of current-based testing, which differs from the historical approach to digital testing (see Figure 2-6). For example, Idd and Iddq or Issq testing are based on the measurement of supply, or return, current instead of on voltage levels.

2.6.2 Current-Based Testing Detection

The observe criteria for a current-based test also differ from the observe criteria for a standard digital voltage-based test. Current-based testing does not require propagation of a fault effect to an observe point; it requires only exercising the fault model and then measuring the current from the power supply. The fault effect observance is the measurement of current, and the detection criteria are the current flow value exceeding some threshold limit or some power rating specification.

The applied limit or threshold is always a "too high" measurement, but based on two different goals: the measure of maximum power use or the measure of quiescent leakage current. The maximum power usage is referred to as Idd current ($P=VI$), and the goal is to not exceed a rated maximum for a very active vector, or set of vectors. The limit may be applied to avoid damaging the silicon or the package, or to meet a specification (for example, to conserve battery usage in a battery-operated device). The other current-based test exploits an interesting effect of CMOS-based design, which is that virtually "no leakage" occurs after the transistors in a gate element have completed switching. The leakage current of a chip in which all the transistors have completed switching should be the sum of the "diode" reverse or quiescent currents for each transistor in the chip (and the fault models used for quiescent current testing are usually the pseudo-stuck-at and toggle models with "quiet vector" selection).

The measurement of current also includes the element of time. If the measurement is taken immediately after a clock edge, then the combinational logic may still be transitioning (which would be when current would be measured for power). However, a current measurement at this time may exceed the threshold for a static or quiescent test. For most static current measurements, the system clocks are stopped, all dynamic devices are disabled, and the measure point follows a waiting period to allow the switching to settle down.

A few extra points need to be made about current-based testing. Current-based testing does not rely on the single-fault assumption, so the detection is sometimes referred to as "massive observability." Since many faults (leakages or switching sources) may be exercised at one time, a fail condition is not easily diagnosed to a specific source. Also, determining the static or quiescent threshold before a silicon process is qualified is very difficult—it still seems to be an experimental process. The amount of current that a chip draws in quiescent state is related to the manufacturing process parameters, the feature geometries, and the number of transistors contained on the die. Current-based testing is another type of testing where subtle faults may exist, and in great number, and they may not be classified as device failures.

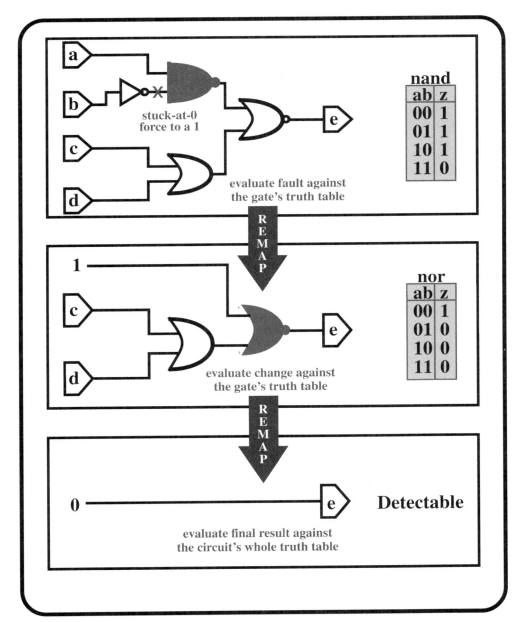

Figure 2-7 Stuck-at Fault Effective Circuit

2.7 Testability and Fault Analysis Methods

2.7.1 Why Conduct ATPG Analysis or Testability Analysis?

Many forms of testability analysis are in common use. Some of the analyses are for the "human understanding" of the fault universe and for understanding the difficulty involved with generating vectors for coverage of certain areas in the circuit. Other analyses are used to help a human or a tool select test points to raise the fault coverage, and some are used to help steer the ATPG engine as it traces through the design description (or to identify the areas that may confound the ATPG engine).

Generally, if the analysis is applied to the design description before ATPG, it is used to get an assessment of the difficulty involved with vector generation or in achieving an expected coverage. If the analysis is applied during design description conditioning (netlist processing), then it is an attempt for the tool to find and add control and observe test points, or to learn better ways to apply and manage the ATPG algorithms. If the analysis is applied during ATPG, it is mostly to steer the tracing algorithms, reduce ATPG tool runtime, increase the likelihood of high fault coverage, or increase the probability of having vector generation success. When the analyses are applied as a post ATPG step, then the reason is to understand the source of "lost" coverage or to help with debug on bad vectors (finding the cause of simulation mismatch).

2.7.2 What Types of Testability Analysis Are Available?

Many types of testability analyses can be applied to the design description during the automated vector generation process. Three of the main ones are "fault effective circuits," "controllability-observability," and "circuit learning."

2.7.3 Fault Effective Circuits

One of the most fundamental analysis techniques is to understand what a "fault" does to a circuit. Understanding the impact that a fault may have on a circuit is sometimes known as "failure modes and effects analysis" (FMEA). For example, in some cases, a combinational stuck-at fault can remap the functional logic to provide a similar function, and an analysis of this similar function may show that a functional test would not easily find the fault. In other cases, when the fault model is applied, the functional logic is radically remapped so that the fault becomes immediately apparent (see Figure 2-7). This type of analysis is sometimes required as a deliverable document in military and government applications, and it is used to provide more information to a technician that must diagnose and debug field failures.

Some analysis methods use circuit remapping to identify the effect that faults have on the circuit, and this analysis identifies whether a fault is easily detectable or not. In some cases, for the DC stuck-at faults, this analysis can be used to identify where test points should be placed to enhance the ability to provide controllability or observability during vector generation. In other cases, the analysis would be used to identify where test points should be added to assist in post-manufacturing debug.

For the example shown in Figure 2-7, this entire circuit would convert to a stuck-at 0, regardless of the input stimulus applied to the circuit. Using this circuit during fault simulation would reduce the compute time involved with vector grading.

2.7.4 Controllability-Observability Analysis

Another analysis that is used to assess the difficulty of exercising, propagating, and detecting fault effects is a controllability and observability analysis. A well known public domain analysis is the SCOAP (Sandia Controllability-Observability Analysis Program). This analysis, and other similar ones, traces through the design description and assigns "Controllability Weights," and "Observability Weights" to the various nets, nodes, gate inputs, and gate outputs. This process is fairly straightforward for combinational logic, but it does get complicated when sequential logic must be included in the analysis.

2.7.4.1 Controllability

For example, a basic controllability analysis may be conducted by placing a weight function of a 1 on every input or direct control point—this includes input pins and points directly reachable by a test point because they have maximum controllability. If the analysis is for the "control-to-logic-1" and if the 1 weighting factor passes through an OR-gate, then the gate output is assigned a 1 as well. If the 1 weighting factor passes through an AND-gate and the other input to the AND-gate is a 1 weighting function as well, then the gate output is the sum of the inputs or a 2 weighting function. The analysis continues with the OR-type gates passing the lowest input weight value to the gate output and the AND-type gates being the sum of the gate input weight values. For the "control-to-logic-0" analysis, the sense is inverted with OR-type gates being the sum of the input weights and the AND-type gates being the passing of the lower input weight. Similar weighting functions can be defined for other gate types such as multiplexors, tristate drivers, and sequential devices. As the weighting function passes through the design description, internal nets take on numbered values. Internal nets with high numerical values have "controllability" problems.

2.7.4.2 Observability

The basic observability analysis is done similarly to the controllability analysis, but it is a backtrace from the circuit outputs and direct observe points. The observability analysis is a measure of being able to propagate a fault effect to an observe point and relies on the controllability weights assigned to the various elements of a design description. To pass a value through an AND-type gate requires placing all the inputs but the one being evaluated to the logic 1 value (logic 1s on all inputs except the one input that will pass a 0 or a 1)—the observability weight of passing the value through this gate, then, is the sum of the "control-to-logic-1" weights on all the inputs but the one being evaluated. Similarly, to pass a value through an OR-type gate requires placing all inputs, except the one being evaluated, to the logic 0 value. The observability weight of passing a value through the OR-type gate is then the sum of the "control-to-logic-0" weights on all the inputs except for the one being evaluated.

The total observability weighting involved with a circuit can be calculated by starting with a weight of 0 at the outputs of the circuit, and then backtracing through the design description while calculating and assigning weights. Again, any internal net with a high numerical value has an observability problem (controlling many gate inputs is required to pass a fault effect to the observe point).

The controllability and observability analyses are generally represented in the design description by assigning a "triplet" of numbers to each gate input and gate output. The "triplet" is written as (C0, C1, O), where C0 is "control-to-logic-0," C1 is "control-to-logic-1," an O is observability.

These analyses have many uses. They may be used to direct the algorithmic tracing functions during ATPG to minimize compute time, and make successful vector generation more likely; or they may be used in conjunction with the non-detected fault list to identify the source logic involved with "thwarting and confounding" the ATPG engine; or they may be used before ATPG to identify the most likely places to place controllability or observability test points.

It must be mentioned that the controllability and observability analyses should be simple fairly straightforward analyses, and not great complicated tracing and assigning functions (which can happen if the algorithms must be modified to handle all manner of complicated circuitry). The reason they should remain simple has to do with computing cost—if the testability analysis becomes too complicated, then it approaches the level of computing cost that ATPG does.

2.7.5 Circuit Learning

There are many different circuit tracing and pattern generation algorithms. The real difference between some pattern generation algorithms (and these differences can lead to higher fault coverage or decreased vector generation tool runtime) has to do with circuit learning. The more the ATPG engine knows about dealing with and operating the circuit, the more easily it can make the correct path tracing or logic value assignment decisions during vector generation. For this reason, the processing of the design description to identify items such as complex gates, multiplexors, tristate drivers, and bus logic, or to add some sort of behavioral description assistance, and to develop from this the best way to trace through this logic, is an almost required portion of many current state-of-the-art ATPG tools.

The reason learning is needed has to do with the way some of the complex logic affects the tracing or logic placement algorithms. For example, a two-input AND-gate needs only a logic 1 to be placed on one of its inputs and this gate can then pass either the logic 1 or logic 0 value through the other input. A stuck-at-one fault on the "A" input requires that the "A" input be driven to a logic 0, and the "B" input must be a logic 1 to pass either the good value of logic 0 or the faulty value of logic 1.

However, a multiplexor is a different story when dealing with faults on the select line. A multiplexor requires that the ATPG algorithm place a logic 0 on one data input, and a logic 1 on the other data input. For example, a regular two-input multiplexor may have a logic 0 applied to the "A" input, and a logic 1 applied to the "B" input by an ATPG tool. During the generation of a complete test set, a stuck-at-one fault is assumed to be on the select line, so a vector is created where the select line is assigned a logic 0 by the ATPG tool. The multiplexor will pass a logic 1 if there is a fault, and a logic 0 if there isn't. The simple act of adding a select line and requiring differing values on the A-input and the B-input makes this gate much more complicated to test algorithmically than the simple AND-gate.

Now imagine an even harder scenario. What if the multiplexor is made of "loose" gates within the design description? To the ATPG algorithms, the logic looks like AND-gates and invertors. The ATPG tool will try to solve the problem with standard "gate tracing" algorithms. However, if a learning analysis were applied, then a collection of gates that implemented the multiplexor function could be identified and the algorithmic testing would know to handle the collection of gates as a multiplexor.

2 • Automatic Test Pattern Generation Fundamentals 57

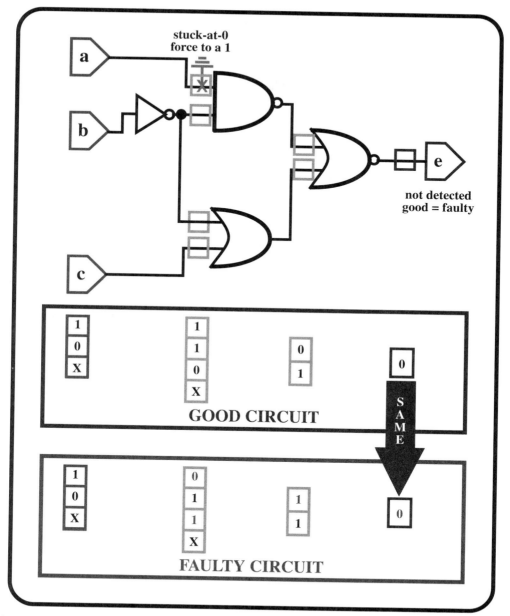

Figure 2-8 Fault Masking

2.8 Fault Masking

2.8.1 Causes and Effects of Fault Masking

Sometimes a problem complicates detection of combinational stuck-at and delay faults when redundant circuitry, reconvergent fanout, or masking circuitry (such as an Error Correction Code logic) exists. This problem manifests itself as a loss of detection, since the fault effect cannot be propagated out to a detection point. Propagation for detection is limited because the effect is blocked by some gate and the output value evident at the detection point is the same as the good circuit value, as shown in Figure 2-8.

A very similar effect is non-detection by ambiguity. In this case, the bad circuit does have the wrong value at the observe point. However, the origin of the bad value is ambiguous. If a fault simulator can't uniquely determine the source of the incorrect expect value, then it does not assign a fault detection value to the target fault since the fault simulator is operating under the "single fault assumption." Also, in the case where the fault simulator assigns a logical X at the observe point, there is a 50% chance that the X will resolve to the correct value on silicon, so some partial credit can be given. Some vector generation tools may keep these types of vectors and label the fault coverage as "non-testable potential."

Closely related to fault masking is the inability to exercise the fault. This occurs when the node that needs to be set to the opposite value to exercise the fault cannot be controlled, and for the same reason as fault masking, a gate blocks the signal because of a conflict between the sensitization for propagating the fault effect and exercising the fault. When a fault cannot be exercised, a controllability problem exists; when the fault cannot be propagated to an observe point, an observability problem exists.

2.8.2 Fault Masking on Various Fault Models

Fault masking is not just a DC combinational stuck-at fault problem. Fault masking can affect all fault models. Redundancy and reconvergent fanout are more devastating to the fault coverage of the more complex fault models. For example, a single redundant node may result in 2 combinational stuck-at faults not being detected, but that same redundancy can altogether eliminate the ability to generate robust path delay vectors for the whole cone of logic, and any other cone of logic that may require this cone of logic in order to successfully complete a path delay test (for example, cones of logic in other time frames—these will be explained later).

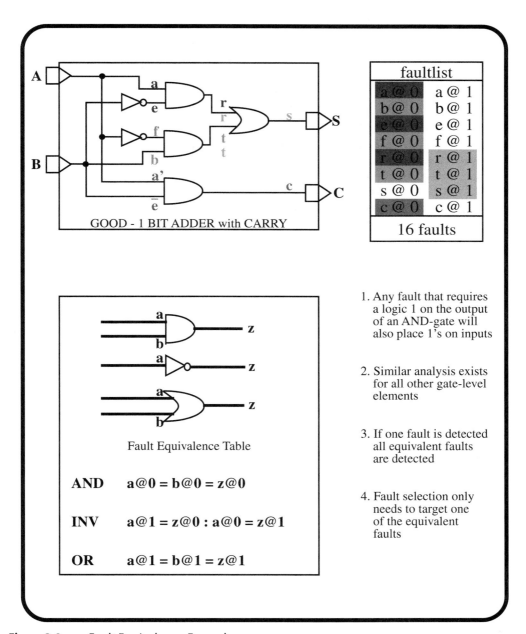

Figure 2-9   Fault Equivalence Example

2.9 Stuck Fault Equivalence

2.9.1 Fault Equivalence Optimization

Another common method used to reduce the tool runtime involved in the ATPG process is to conduct fault dropping based on fault equivalence assessment (see Figure 2-9). Stuck fault equivalence allows fault collapsing to be accomplished. Fault equivalence is a static analysis that allows the gate output faults to be used to eliminate gate input faults from the faultlist (to collapse the input faults into the output fault). If the fault selection process lists the gate output fault first, then many of the gate input faults do not have to be included in the list, and they will not have to undergo vector generation.

Fault equivalence is also commutative in that backtracing from some gate outputs will backtrace through several gates, and so one fundamental fault may have many equivalent faults back through the circuit. An analysis strategy that optimizes the ability to drop faults from consideration by fault equivalence is to begin the fault enumeration analysis nearest to the circuit or system observe points (for example, output pins, scan flip-flops, and test points).

2.9.2 Fault Equivalence Side Effects

Fault equivalence may reduce the number of faults to be considered and may reduce ATPG time, but it may also lead to some loss of diagnosis. The vectors created for a fundamental fault near an observe point will automatically exercise and detect all of the equivalent faults (for the proper logic sense). However, a single vector now has several faults associated with it (a fact that is actually very good for fault denseness and fault efficiency).

However, if the ATPG tool automatically conducts fault equivalency on a circuit to eliminate faults from the faultlist, then the ability to use that tool's faultlist to assist in diagnostics vector development is questionable. For example, one method used to isolate the source of several failures is diagnostic fault simulation—several failing vectors are fault simulated with fault dropping turned off. If a single fault in the circuit can be the source of all the failures, then that fault is the most likely candidate. Sometimes this process requires generating more vectors around a suspected fault—if the ATPG tool presents only the fundamental fault in the faultlist and not the equivalent faults, then generating vectors for the diagnostic fault simulation may not be successful.

2 • Automatic Test Pattern Generation Fundamentals

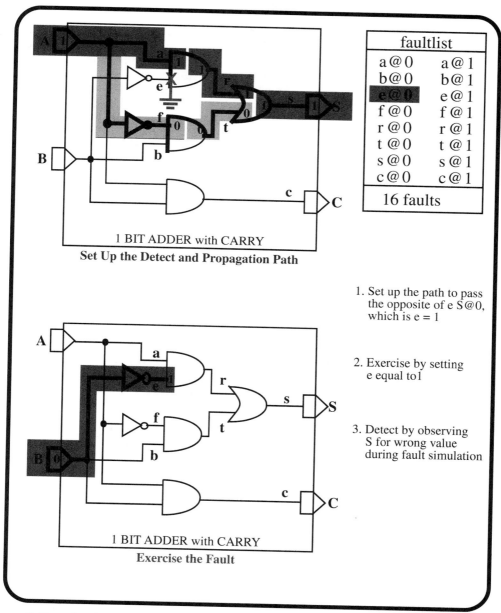

Figure 2-10 Stuck-At Fault ATPG

2.10 Stuck-At ATPG

An Automatic Test Pattern Generation (ATPG) tool is used to solve the problem of structural vector generation. All of the most common ATPG tools are designed to create vectors to exercise and detect the combinational stuck-at fault, and to convert the fault into a failure. This process generally occurs in at least two basic algorithmic steps. The first step, after a fault has been selected, is to exercise the fault. The second step is to sensitize a detection path from the fault location to the observe point. The actual algorithmic steps applied by any specific tool may number more than these.

For example, the very first step may be to "intelligently" select the fault as close to an observe point as possible, and then to exercise the fault location by tracing backwards from the fault location to a control point that will allow the fault to be exercised, and to also trace backward from the other inputs to each gate in this path to provide these control points. The next step would be to trace from the fault location forward to the observe point to establish the path the fault effect will take and to trace from the inputs of the gates in the observe path backward to control points to establish the proper sense of the path. As can be seen in Figure 2-10, four separate tracing functions are applied to conduct this example: 1) backward from the fault to a control point; 2) backward from the fault to control points for each gate in the path between the fault and its control point; 3) forward from the fault location to the observe point; and 4) backward from the observe point to control points for each gate in the path between the fault and the observe point. Even more tracing strategies (subroutines) may be applied if any of these traces are unsuccessful or in conflict with each other.

Note that limits may be placed on the tracing processes to manage the time involved with generating a vector. Two of the most common limits are the "abort limit" and the "backtrack limit." These functions can set bounds on the time the algorithm spends trying to generate a vector, or can actually place a target value for how many nets, nodes, and circuit elements are involved in the tracing functions.

Many algorithms have been developed to facilitate the generation of combinational stuck-at vectors. Some of the most common ones are the D algorithm and PODEM. Since this is a light tutorial, specific algorithms and nuances will not be discussed other than to state that the big difference in the algorithms is the order that the algorithmic steps are applied (tracing the detection propagation path, tracing the detection enable path, tracing the fault exercising path, and tracing the fault exercising enable path), the directions used for tracing (forward or backward through the design description), and the extra circuit information needed to increase the likelihood of success.

2.10.1 Fault Selection

The very first step in the automatic creation of a structural vector is the identification of the fault. This is generally done by listing all gate inputs and gate outputs as both the stuck-at one and stuck-at 0 faults. Sometimes the nets used as interconnects are included in the analysis (depending on whether the design description includes the concept of "wires" or "nets" as opposed to just gate connections). The total number of objects included as faults does affect the tool runtime, the size of the computer image, and the final measurement or metric.

2 • Automatic Test Pattern Generation Fundamentals

Once the faults have been "enumerated" and listed, the fault selection process can begin. However, the order of the listing is important and can affect the algorithmic runtime. For example, an order can be established that can be alphabetical, hierarchical, random, based on the order the design description is processed, or some other optimized form. The fault listing can also be optimized or minimized at this point by using analyses such as fault equivalency.

As was stated earlier, an intelligent fault selection process can reduce tool runtime significantly. Forward tracing through a gate-level netlist is much harder than backtracing through a netlist (place your finger in the middle of a field of logic and trace forward to any of several identified observation points, given that the circuit has widespread fanout—or place your finger on the observation point and trace back to any given gate—backward is easier). To this end, selecting faults on the list to minimize the analysis involved with the more difficult part of the algorithms results in a more optimal tool runtime. In this case, selecting the input to the observe point as the fault results in the minimum amount of forward tracing and eliminates one algorithmic step, the tracing of the detection propagation path back to the control point. This optimized fault selection step reduces the algorithm to just the two steps of exercising of the fault and enabling the exercise path. When the produced vector is fault simulated, many of the "equivalent" faults in the "exercise" and "control" paths are also detected.

2.10.2 Exercising the Fault

The most fundamental operation of the ATPG tool is to calculate how to exercise the fault. This is done by solving from the fault backwards to a set of stimulus or control points. Values are placed on all control points necessary to exercise the fault.

2.10.3 Detect Path Sensitization

One of the more difficult parts of vector generation is detection path sensitization, because forward propagation of the fault effect to the observation point is the most difficult step. It takes more processing power and time to forward solve from the fault to a detect point. Part of this is because more than one detection point may be available. The limits on processing established by some tools favor finding the "nearest" detection point.

The tool will assume that the faulty node is driven to the opposite value of its stuck-at and will then attempt to place all the necessary values on the gates, from the fault to the most accessible observe point, so that the fault effect will be propagated and detected (for example, a stuck-at 1 on a gate input will result in the ATPG tool attempting to propagate the logic 0 from this gate input). This approach can be viewed two different ways: either the path for the good value is created in an exclusive manner (always choosing the unique path through the gates), or a specific path is created that will pass the assumed bad value. Generally, the passing of the good value is preferred and a unique path is crafted by enabling an AND or NAND by placing a logic 1 on the off-path inputs, and by enabling an OR or NOR by placing a logic 0 on the off-path inputs. This will allow only the passage of the good circuit value, and any other value observed at the observe point will have detected the fault.

A problem that arises during the various tracing steps is conflicts in "sensitizing" paths. For example, if the exercising of a fault on some circuit element requires placing several other gate inputs to certain logic states and then creating the detection path to the nearest observe point also requires placing some of the same gate inputs to the opposite logic states, then a conflict occurs. A solution may require finding an alternate set of pathways to exercise the fault or finding an alternate set of pathways to propagate the fault effect to an observe point.

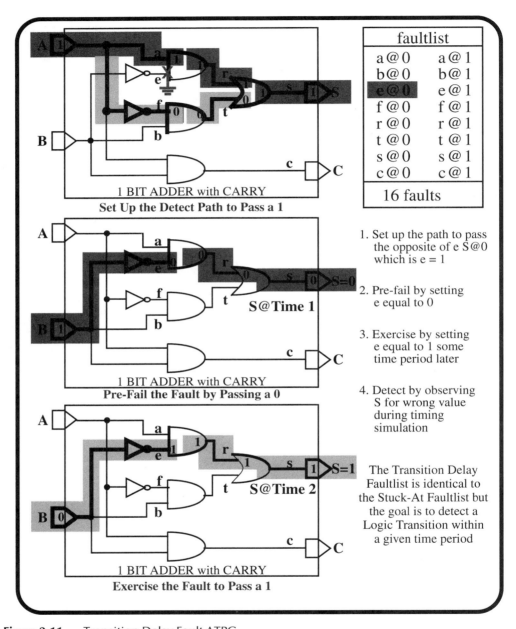

Figure 2-11 Transition Delay Fault ATPG

2.11 Transition Delay Fault ATPG

2.11.1 Using ATPG with Transition Delay Faults

The transition delay fault model (shown in Figure 2-11) is used to verify the timing structure of a circuit. Timing is considered by launching "logic transitions" or "vector pairs" into circuit elements and then assessing the propagation delay timing difference between the two different logic values. Although the test may seem completely different from the stuck-at fault type of test, the algorithms and assumptions made are almost identical to stuck-at fault ATPG. The only significant difference, which is what adds timing to the analysis, is to require that the output of the gate (or the detect point) be set to the fail value first. This change allows the same structural elements operated on by the stuck-at fault (gate inputs, gate outputs, interconnect nets) to be assessed for timing. Because of this, the transition delay fault is also known as the "gate delay" fault.

For example, if an AND-gate is to be evaluated for timing, a set of tests would be crafted to test the reaction of the gate inputs to a rising signal and a falling signal. This process would require four tests (0->1 and 1->0 on the A-input, and 0->1/1->0 on the B-input), and the gate output would be verified as part of the gate input tests (by fault equivalence). To apply a slow-to-rise transition test to the A-input would require placing a logic 1 on the B-input and a logic 0 on the A-input during the first time frame, and then at some time later changing the logic value on the A-input to a logic 1 while holding the B-input at a logic 1. This action has the effect of launching the 0->1 vector pair on the A-input, which would evaluate the A-input and the gate output for the slow-to-rise transition delay fault. Note that the target gate is situated within the circuit, so all the fault exercising, fault propagation, and path sensitization tracing steps involved with relating the test actions to the control and observe points as described with the static stuck-at fault ATPG must be applied.

The fault selection process, in reality, is identical to the stuck-at process (a finite number of circuit elements are enumerated and are related to two faults each—they are just viewed as slow-to-rise and slow-to-fall instead of stuck-at 0 and stuck-at 1). Also, the detect path sensitization can also be done exactly as the stuck-at process (and fault equivalence holds in a limited manner). The fault exercising step is different. The gate being evaluated for the fault must be set to the fail value first, and this fail value must be allowed to propagate to the detect point and be observed. The gate is then set to the pass value (a logic transition has been launched) and the detect point is observed at some given time period later. If the pass value is evident, then the circuit is failure-free. Note that there may still be a fault if the pass value is observed, but the delay contributed is less than the time target applied as the standard for measurement. If the fail value is observed, then the transition did not propagate in the required time, and a slow-to-rise, slow-to-fall, or signal propagation delay is assumed to have caused the failure.

Also note that adding the extra step of applying both the logic 1 and logic 0 to the identified fault as part of the exercise step makes the transition delay test a more difficult or more compute-intensive problem. For this reason, transition delay fault coverage does not usually match the stuck-at fault coverage, even though both may be working on the same enumerated list with the same number of faults. In that same vein, however, stuck-at fault coverage can be supplanted with transition delay fault coverage for the same faults—the worst-case fault for a transition test to detect is an infinitely slow delay fault, which is a stuck-at fault.

2.11.2 Transition Delay Is a Gross Delay Fault

As mentioned earlier, the algorithms applied to solve the fault detect path sensitization and fault exercising steps can be very similar to the stuck-at vector generation analysis. The work function of these algorithms is usually to find the shortest path to exercise the fault, and also to find the shortest path to drive the fault effect to the detect point. This generally results in a "minimum" path (timing) vector that more relies on the targeted fault being a significant delay adder, or a "gross" delay fault. The resultant vector (in the absence of a delay fault) can be viewed as being more suited for verification of a capture flip-flop's "hold time," rather than the maximum delay verification of the flip-flop's "setup time."

2 • Automatic Test Pattern Generation Fundamentals

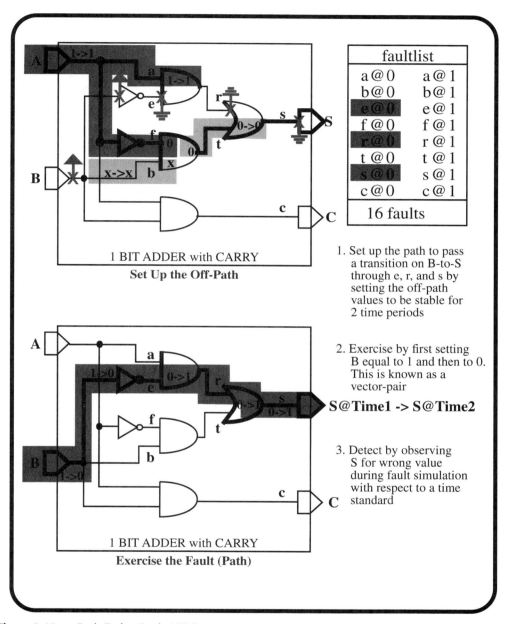

Figure 2-12 Path Delay Fault ATPG

2.12 Path Delay Fault ATPG

2.12.1 Path Delay ATPG

The path delay fault model is different from the transition delay fault model and the stuck-at analysis in many ways. Whereas the transition delay vector is attempting to exercise and detect the "gross" delay fault, the path delay vector is more trying to exercise and detect the "subtle" delay fault on a described circuit path (see Figure 2-12).

Another significant difference between the transition delay or stuck-at fault analysis and the path delay analysis is the fault set. For both the transition delay fault model and the stuck-at fault model, there is a finite number of elements (gate inputs, gate outputs, and interconnect nets). The path delay fault is based on the concept of a path which is a specific collection of gate inputs, gate outputs, and interconnect nets. Another way of viewing this is to say that the path delay fault is the "stack up" of a collection of transition gate delay faults. The number of paths within a design description may be finite, but it is much larger than the number of elements, and is generally larger than is easily computable (it is much like the probability of winning a lottery—50 elements taken in groups of 6 result in over 15 million combinations—and a netlist may start with millions of elements; even reducing the listing based on circuit dependencies still results in more paths than can be economically tested on a tester). Therefore, the path selection process (which is the fault selection process) must be constrained or limited in some manner.

However different the fault models may be, the path delay fault is similar to the transition delay fault in that the analysis is done in "two time frames" with three actions. The first time frame requires establishing the fail value and launching a transition. The second time frame captures the effect of the transition.

2.12.2 Robust Fault Detection

As was described in the transition delay section, the gate inputs that are not part of the transition test must be held to an enabling value throughout the test (for example, a logic 1 on AND-gates and a logic 0 on OR-gates). In the case of path delay, there are many enabling gate inputs since the "effective fault model" is a description of all the gates, nets, and nodes involved with a path. These enabling gate inputs are generally referred to as "off-path" signals. If the ATPG is successful when the "off-path" signal values are held constant for the two time periods involved in the test and if the detected result is due only to the launched transition, then the vector, or fault detection is labeled as *robust*. Sometimes it is not possible to hold the off path values stable for two time periods, or it is not possible to determine whether the pass value was from the launched transition or whether it was from a redundant "sneak" timing path (a path that may provide a bypass around the path being tested). When this type of vector is made, the fault detection is labeled *non-robust*. There are other detection sub-categories that have to do with whether the vector is hazard-free, hazardous, redundant, and so on, but the best vector to generate and keep is the robust vector (even though this is the hardest vector to generate).

As was described about the transition delay vectors and how they can supplant the stuck-at vectors for the same fault coverage (and how they are more difficult to generate), the path delay vectors and fault coverage can supplant both the stuck-at and the transition delay vectors. However, in most cases, the path delay vector is the most difficult vector to generate, and converting the various path elements to transition delay and stuck-at fault coverage results in less overall coverage—so the path delay can't be used to replace all the transition and stuck vectors.

2.12.3 The Path Delay Design Description

Many ATPG tools receive their "paths" from alternate sources. One of the restrictions involved with the path delay fault model is that the design description that the ATPG operates on must closely resemble the design description from which the critical timing paths are taken. The timing analysis tool will operate on a design description (a gate-level netlist), and it will create the list of critical paths, and the critical path descriptions used for ATPG. When these paths are fed to the ATPG tool, the ATPG tool must be able to find the "exact" paths. Just denoting the starting, or transition, point and the ending, or capture, point is not sufficient—hundreds of paths may exist between the two points. The key is to generate the vectors on the "critical paths," not the ones "next to" or "near" the critical path. The brute-force method of generating vectors for "all" paths around a critical path may not be economical in both vector generation time and in the number of vectors produced.

The "exactness" problem makes the link between the timing analysis tool and the ATPG engine very sensitive to netlist formatting problems such as "underscores" versus "slashes" versus "dots," and "hierarchical" names versus "flattened" names. If the two design descriptions are different, then some form of path description translation must exist to convert the path descriptions from the timing analysis tool to the path description format for the ATPG tool, and this translation must incorporate all the little "nit-picky" differences.

A similar path description problem exists when "custom" logic is in the design description. Custom, or hand-layout, logics are usually transistor-level design elements that are represented in the design description with an equivalent circuit. The critical timing paths for this logic are measured with a transistor-level or switch-level simulator. When the timing is annotated to the gate-level design description, it is usually just a timing associated with an element connection. However, for the ATPG tool to create a vector and to sensitize the correct critical path, the custom logic block must be delivered with not only the connection timing value, but also the logical values that must be applied to the other connections (in effect, the timing vector created by the transistor-level tool must be transferred in some manner to the gate-level tool).

2.12.4 Path Enumeration

Some ATPG tools do not receive their paths from a timing analysis tool; they just create them by tracing through the design description contained within the ATPG tool. This gives the ATPG engine paths to operate on, but no "criticality" or timing information is related to these paths. This type of "path enumeration" must be filtered after the vectors are generated to identify the vectors involved with the critical paths (if the vector set must be limited or reduced from the full set generated on all possible paths).

Some of the more aggressive modern tools are combining static timing analysis with the ATPG tool. This combination eases the selection of paths, solves the non-identical netlist problem, and eases the complexity of the whole methodology. The caution here is that the ATPG tool's internal timing analysis is generally not the same tool as the "official" timing analyzer that the design may be signed-off against—the two tools may differ in the assessment of the most critical paths (however, history has shown that even though individual tools may uniquely rate individual critical paths differently, the top 10% of critical paths selected by different tools generally have the same critical paths in the overall collection).

70 — Design-For-Test for Digital IC's and Embedded Core Systems

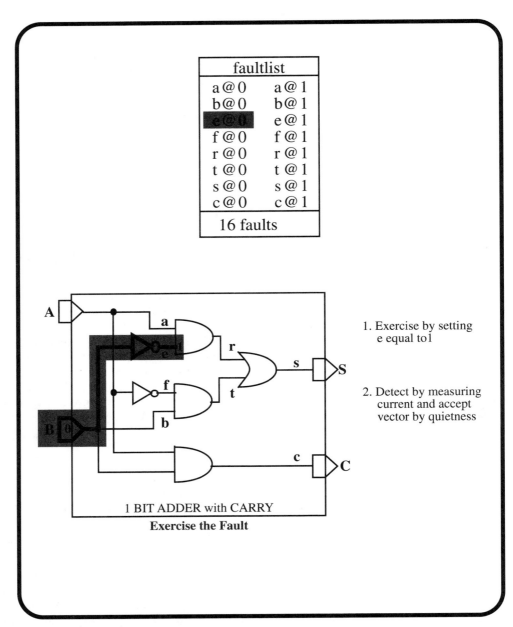

Figure 2-13 Current Fault ATPG

2.13 Current-Based Fault ATPG

2.13.1 Current-Based ATPG Algorithms

Current-based testing, illustrated in Figure 2-13, is a completely different paradigm from voltage-based (logic level-based) testing. The goals of current-based testing are different as well. The measurable is the amount of current being supplied by (or returned to) the tester's power supply, and the measurement's fail criteria is an amount of current larger than some established threshold. Because the measurement is the current applied at the chip-level, the fault model can't be restricted to the single-fault assumption. With structural-based logic level testing, linking an erroneous logic response to the logic that created it is fairly straightforward under the single-fault assumption. However, an excessive current leakage can have its source anywhere in the logic connected to the power supply being measured.

Because the single fault assumption does not hold, current-based testing, in the ideal sense, needs only the "exercise-the-fault" step (and to exercise many faults) and needs to ensure that no driven contention or active (intentional) leakage exists in the circuit. The measurement of the current supplied or drawn at the power supply can then be used as a "massive observability" event.

However, in the ATPG world, it is much simpler to use the stuck-at model and existing vector generation tools, and to apply the extra step of simulation to assess the quietness of the circuit (that no driven contention exists, or that any dynamic devices are not activated such as pullups and pulldowns). In fact, some current-based testing tools are not ATPG tools, but are fault-grading tools that advocate the use of existing functional or scan vectors filtered for quietness. ATPG tools, on the other hand, generally support the generation of quiet vectors based on the stuck-at, pseudo-stuck-at, or toggle fault models. For example, using the stuck-at model and creating a voltage test, and then conducting a current measurement, and also conducting the voltage measurement afterwards, leads to a multiple clue test—but the failure of just the current measurement is not conclusive.

It is more efficient to generate current-measurement vectors by using the exercise-only part of the ATPG algorithm. For example, a toggle fault model attempts to apply the logic 1 and logic 0 value to as many nets and gate connections in the design (to verify that every net can be driven to both logic levels), but the fault model does not try to drive the fault effect to an observe point. In just a few vectors, the toggle fault coverage can be very near its maximum value (for example, 98%).

The application of current-based testing is less mature than voltage-based testing. The tools and the fault models that are publicly available are still fairly rudimentary. For example, there are no real comprehensive defect-based fault models (fault models based on understanding the behavior of defects that create current leakage) to assist with diagnosis and isolation. Also, no good analysis tools are available to help establish the threshold value (which is related to the process, process geometry, and the number of transistors on the target device).

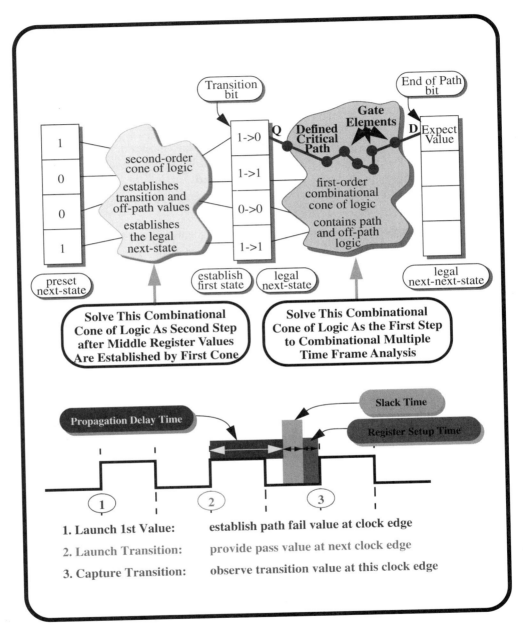

Figure 2-14 Two-Time Frame ATPG

2.14 Combinational versus Sequential ATPG

Up to now, all the information presented has treated the design descriptions as if it comprised only combinational logic. In reality, ATPG is applied to both combinational-only and mixed combinational-sequential circuits. Even though the combinational ATPG problem is much easier to solve, some logic structures can't be modeled or dealt with as a combinational-only circuit. Structures that require solutions in "multiple time frames" can be dealt with as either a "sequential" problem or a "multiple time frame combinational" problem.

Sequential ATPG is the application of the tracing and logic value selection algorithms against a design description that includes sequential elements. This requires that the algorithms be expanded to understand how to operate the sequential elements. Sequential elements may be the different types of flip-flops (SR, JK, T, and D), latches, and even some complex memory structures.

Combinational ATPG for multiple time frames uses the same combinational algorithms as for ordinary combinational ATPG, but makes some simplifying assumptions on how the sequential devices work so that it can leave them out of the analysis. For example, all sequential devices may be assumed to capture data identically when the clock is toggled. Combinational-only vector generation is usually applied to a design description that has been rendered effectively combinational by the application of a structural testability methodology such as full-scan.

2.14.1 Multiple Cycle Sequential Test Pattern Generation

Sequential vector generation is needed or required only if the design description can't be dealt with in the strict combinational sense (there is a need to pass a logic value through a sequential element). Sequential vector generators have also been referred to as "partial scan" vector generators. If the circuit description is "non-scanned," "mostly-scanned," or "partially-scanned," then the vector generator needs to solve the problem of propagating logic values through various time frames (each clock cycle is a different time frame). A sequential ATPG tool, for example, must apply logic values to the input pins of the chip, and must then propagate these values through the circuit to the output pins, and it must operate the sequential elements within the chip as needed.

The ability of the ATPG tool to operate the sequential elements, and how the tool operates the sequential elements, is what can make a discriminating difference between different ATPG tools. A very comprehensive tool includes algorithms to understand and prioritize the selection of updating sequential elements by understanding the set/reset logic, the ability to hold data with an enable or denying the clock, and the normal operation of placing data on the data input port and applying a clock signal. Tools that are "partial scan" must understand when to view some of the sequential elements as "not being there" for the scan load process, and when to view the sequential element as being part of a non-scan sequential operation (some of the sequential elements exist and don't exist, depending on whether shifting or sampling is the current operation).

Because of the added complexity of the sequential elements, sequential ATPG uses much more computer memory and is generally much slower than combinational ATPG, even if the circuit is rendered combinational-only by making it full-scan (the applied algorithms are still based on the existence of sequential elements even if none exist).

2.14.2 Multiple Time Frame Combinational ATPG

Note that the majority of sequential or multiple time frame vector generators are simply used for single stuck-at faults in the presence of sequential elements. However, having the ability to handle multiple time frames enables these tools to process the path delay and transition delay fault models as well. The path delay and transition delay fault models both require vector pairs (vectors applied in two consecutive time frames). This requirement allows a fail condition to be applied initially and for a transition to be applied with reference to a timing event (such as a clock rising edge). A time event or clock cycle later, the result of the transition is captured and observed. The ability to do so with a combinational-only analysis requires the ATPG tool to support "time frame analysis" or "combinational sequential depth analysis." Including the ability, in a combinational-only tool, to deal with sequential analysis is actually allowing a vector to be created in one combinational cone of logic, and then calculating the vector needed in the cone of logic that feeds the first cone of logic where the first and second cones of logic are separated by sequential elements (see Figure 2-14). This has the effect of generating the next vector back in time for the first combinational cone of logic with respect to sequential devices not explicitly existing in the ATPG analysis. This process is done by using a method known as *multiple time frame analysis*.

The only reason that a combinational ATPG tool should support a pseudo-sequential ability is for the "mostly scan" and full-scan "delay" vector generation. Mostly scan is a methodology where the design is rendered combinational-only, except for a few places where there are non-scanned devices (as opposed to a true partial-scan methodology, where only a few scannable elements are added to a largely non-scanned design to enhance coverage).

As was described in the path delay section, certain inputs must stay constant for two time frames (1->1 or 0->0), while other inputs must transition (1->0, 0->1) to meet the criteria of robust testing. Since the fault effect is measured against a time standard or time frame, testing requires at least 2 time frames, so a combinational or sequential depth of at least 2 is required.

In the scan ATPG world, the two-cycle analysis is based on register-to-register clock-based data transfers, meaning that an initial state is generated to exercise the path in time frame one (to the fail state), and the fail value is also placed in the capture point. The second time frame (one step back in time in generation order) requires the analysis and generation of a vector that will hold the off-path propagation elements stable, and that will launch the transition on the active path.

When this is played back on the tester, the scanned-in vector will, in the first-order cone of logic:

1. establish the fail at the capture point,
2. set the path up to fail at the launch point,

   and in the second-order cone of logic, will:
3. launch the transition on the next clock edge.

When the next clock edge occurs, all logic values remain the same, except that the transition bit is launched. On the subsequent clock edge, the result of the transition is captured in the observe point. If the path (or a gate in the path) is slow, then the fail value will remain. Another way of saying this is that the ATPG must solve the problem from the end (capture) to the beginning (initial state), while the tester will play it back in reverse (initial state to capture).

2.14.3 Two-Time-Frame ATPG Limitations

A problem with "two-time-frame" ATPG is that vector generation may be unsuccessful because so much more logic is involved. For most ATPG, only one combinational cone of logic is analyzed at a time. Redundancies and reconvergence may block some ATPG within this cone, but not to a great extent. For "two-time-frame" ATPG, redundancies and reconvergence in the second-order cone of logic can thwart successful ATPG in either the second-order cone or the first-order cone. However, one school of thought states, "If a valid state-nextstate cannot be achieved, then the fault is on a false path." History has shown that two-time-frame ATPG successfully generates vectors on very few of the total number of paths in most designs, and this fact is mostly attributed to false path content.

2.14.4 Cycle-Based ATPG Limitations

The cycle-based ATPG is a form of vector generation where a runtime optimization has been applied to an ATPG tool that only assesses changes to the circuit on a defined edge of the clock (for example, the positive edge of the system clock). The runtime optimization of cycle-based ATPG is very similar to the difference between the two major simulation engines, event-based versus cycle-based.

This means that any anomalous behavior that occurs between the clock cycle may not be understood. For example, factoring the clock as data can make a gate into a pulse generator—since the tool does not base any simulation on the negative edge of the clock, then the reaction of the circuit to the negative edge will be missed and can result in incorrect vectors.

Cycle-based ATPG may also result in a loss of accuracy, since dealing with only the "positive edge-to-positive edge" consideration of a clock has limitations in dealing with the negative edge events associated with the "entrance to the last shift" and other events that rely on the negative edge of the clock. Cycle-based ATPG also has difficulty with timing considerations in that some paths may be designed to be only half-cycle paths (from positive edge-to-negative edge, or from negative edge-to-positive edge). The other problem is with multi-cycle paths that may be one-and-a-half-cycle paths.

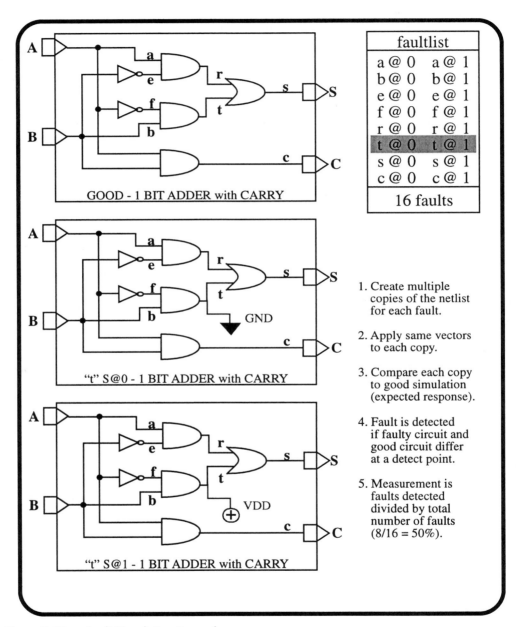

Figure 2-15 Fault Simulation Example

2 • Automatic Test Pattern Generation Fundamentals

2.15   Vector Simulation

2.15.1   Fault Simulation

Fault simulation, fault grading, or vector grading as it is also called, is the process of applying a set of vectors to a structural (netlist) description of a design and determining how many and which faults are detected out of the total set of available faults. The measurement used by simulators for fault grading is usually displayed as a percentage of faults: 90 faults detected, out of a total of 100 faults possible in the circuit, is 90% fault detection for the applied vector set. This is known as grading a vector set (establishing how good a vector set is).

To conduct a fault simulation requires a structural netlist description of the design, a vector or set of vectors (logic 1s and 0s applied at the inputs to the design), a list of faults (usually generated as an option by the simulator itself—or a set defined by the ATPG tool), and a library description of the gate level constructs.

The most common type of fault simulation is concurrent fault simulation. This method applies the vectors to many copies of the netlist at the same time. Each copy contains one or more faults. Each of these simulations is run concurrently with a good circuit simulation (see Figure 2-15). If a difference is evident at the legal observation points between the good circuit simulation and any faulty circuit simulation, then the fault is listed as detected and the vector (or time stamp), and detection point is logged (because a fault can be detected by the same vector at multiple output or detect points).

2.15.2   Simulation for Manufacturing Test

The simulation of a circuit is sometimes required to create the manufacturing test program. Earlier it was stated that the three requirements for testing are (1) a known stimulus, (2) a circuit in a known state, and (3) a known expected response. The vectors (the known stimulus) can come from a designer or from an ATPG type of algorithm. In order to physically test a real circuit on a tester, the expected response is needed. This comes from the good circuit simulation that is run independently on the vectors or from the good circuit simulation that is run within the ATPG process (and is written out with the vector by the ATPG tool).

It is also the simulator's task to establish the known state based on the applied vector. If the design is full-scan, then the requirement for known state does not exist because no effective state holding devices are contained within the design (so a new and full state is installed by the tester each time a new vector is scanned in). However, if the design contains uninitialized non-scan sequential logic, then areas of the circuit may be ambiguous at the beginning of, or during, the vector application.

If the design is largely "sequential," then there is the problem of establishing a new known state with each vector, or of simulating and concatenating patterns so that the initial state for each new pattern is the old state from the last pattern. If the vectors are created by a sequential ATPG tool, you should understand what method each specific tool supports. A sequential tool may create independent patterns that contain the full reset sequence for each pattern (it doesn't care what the initial state of the device is—it creates a new one); the tool may create patterns that depend on the state left by the previous pattern—so the vector sets can be used only in the order that they are created; or if the vectors are to be used out of order, then they must be re-simulated to create new expected responses, and their fault coverage may change since the initial conditions for each vector have changed.

One of the points to make here, though, is that the fault simulation, whether done separately or within the ATPG tool, needs to emulate the limitations of the tester. If embedded circuit test points exist but are not accessible by the manufacturing tester (for example, ebeam probe points on silicon), then they should not be used to calculate fault coverage for the manufacturing test. Thus the ATPG tool or simulator needs to be constrained to match the real actions and abilities of the target test platform.

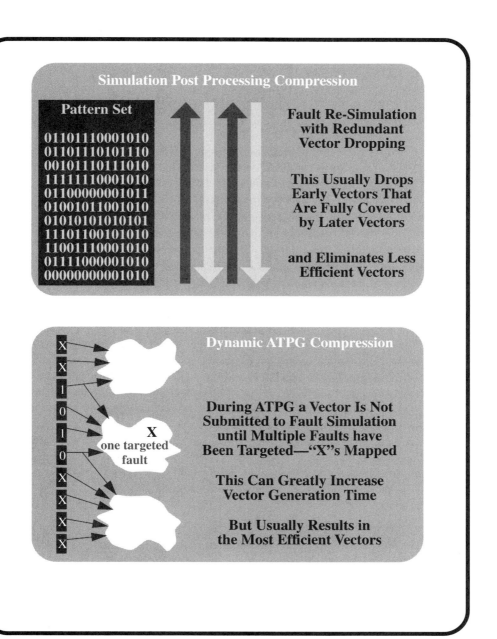

Figure 2-16 Vector Compression and Compaction

2.16 ATPG Vectors

2.16.1 Vector Formats

The ultimate result of the ATPG process is vectors. However, the job isn't over yet. Once the vectors exist within the ATPG tool's memory structure, many questions still need to be answered. To begin with, an ATPG tool can use multiple vector data output formats to present its vector data. One of the main formats is the ATPG tool's own binary format, which is usually the most compressed format (takes up the least disk space) when written out. The binary format is important in that it provides an "archival" vector set that can usually be read back into the ATPG tool for various purposes: diagnostic evaluation on failing vectors; re-output with a different data format; re-fault simulation; or expansion by adding (concatenation of) more ATPG vectors.

Also, multiple formats can be written out to support testers and simulators. Some are based on industry standards such as Verilog and WGL (pronounced "wiggle"); others may be based on the tool providers ASCII format; and some are specific to the target tester.

When vectors are written out, some thought should be given to archival copies and configuration management. In most cases of using the ATPG process, several vector formats may be written out for various purposes and to interface with various tools and testers. However, vector files are "huge." So the question is, "Which vector sets in which formats should be saved?" Some organizations use "configuration management tools" to save the specific version of the ATPG tool, the specific version of the design description, and a copy of all the scripts that were used to generate a specific set of vectors. These allow the vectors to be re-created at any time. Many organizations also require keeping the final translated version of the vectors that are used in the test program—or are used by the test program builder. And, as mentioned before, the tool's own binary or ASCII format is kept in case the vectors must be read back into the tool for re-formatting or for diagnostic purposes.

The reuse of "cores" from a "parts bin" or from a "technology shelf" has also resulted in keeping completed or "packaged" vector sets as well. These vectors must be available in all forms that a customer using the core would require. In any case, the vectors should be stored in a compressed format (for example, ZCS in the UNIX environment).

2.16.2 Vector Compaction and Compression

When ATPG is used for vector generation, a simple systematic method is applied to decrease the number of total vectors required for high fault coverage. This method is to apply "random" or "pseudo-random" vectors first to eliminate all the "easily detected" faults (this also reduces ATPG time). When the coverage curve of random vectors starts to flatten out (all the low-hanging fruit has been picked), then the deterministic vector generation begins. Deterministic vector generation creates vectors targeting a fault, and then this vector is fault-simulated to remove the other faults it detects from the faultlist. As vector generation continues, the faultlist gets smaller and smaller. At some point in time, vector generation seems to be producing one vector per fault because the faultlist itself is composed of only "hard to detect" faults. Experienced DFT and test engineers have been known to quote the old adage, "Getting the last 2% takes about as many vectors (and ATPG runtime) as the first 98%."

One of the early problems with ATPG is that the deterministic vector generation for stuck-at faults did produce many more vectors than most test budgets allowed (exceeded tester time or tester vector memory). The economics of this situation resulted in several methods being developed to reduce the number of vectors (or from another point of view, to make the delivered vectors more efficient in a fault coverage per cycle). One brute-force method was simply to take the delivered pattern set after ATPG and to combine several vectors together by mapping the "X-space" (this works easily only for combinational-only vector sets). This reduces the number of vectors by combining them together. However, this has a limit and is more successful for combinational vectors arranged in "fewer and longer" scan chains.

Another post-processing method is to re-fault simulate the vectors without fault dropping. Any redundant vector is dropped. Redundant vectors exist because the ATPG process naturally removes faults from the faultlist, so vectors generated later in the process may be better vectors, but this possibility is not known, since faults already detected are dropped from consideration. Re-fault simulation without dropping does this analysis after the vectors are generated. It is sometimes useful to conduct the simulation in several vector sequence orders: forward, reverse, and random.

One of the most effective methods is "dynamic" vector compression (see Figure 2-16). This type of compression is the continual targeting of faults until the generated vector has some percentage of bits mapped to active values rather than "X"s before submitting the vector to the fault simulator (a deterministic vector is made that targets more than one fault). For example, scan chains are fully "packed" or "mapped" this way to generate the fewest number of vectors as possible.

Other dynamic methods can be applied such as conducting the fault simulation, in conjunction with ATPG, without fault dropping, and eliminating vectors as more efficient ones are being generated.

Note that most vector compression methods are applied to combinational-only or full-scan vector sets. Conducting static or post-processing compression methods is much harder when multiple time frames (sequentiality) are involved. Multiple time frames or sequential vectors are more easily compressed by the non-dropping fault simulation method used to eliminate redundant vectors.

Also, the use of static or dynamic vector compression or compaction methods reduces the number of vectors, but should not reduce the number of faults detected. Because of this fact, compressed vectors are very active and may consume more power.

Finally, if vector compression and optimization eliminate vectors with redundancy (eliminating vectors that achieve coverage of the same faults), then vectors can be created that do not contain very good diagnostic information. The reason is that individual vectors cover many faults, but not a lot of overlap is involved with the fault coverage. For example, if a fault is only exercised and detected once and that vector covers many equivalent faults, then an inherent ambiguity exists when that vector fails—no other "multiple clue" information provides further isolation.

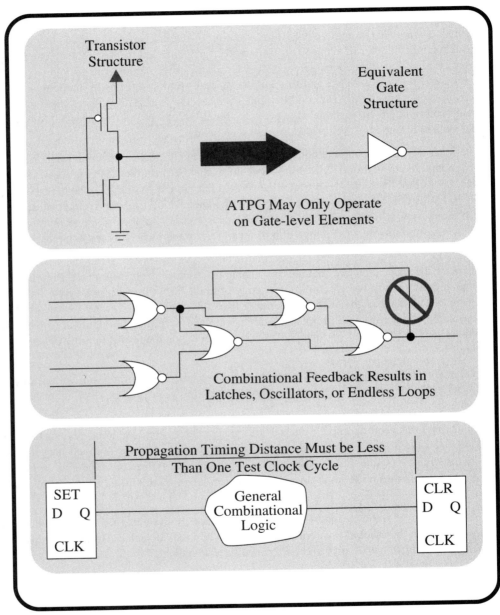

Figure 2-17 Some Example Design Rules for ATPG Support

2 • Automatic Test Pattern Generation Fundamentals

2.17 ATPG-Based Design Rules

2.17.1 The ATPG Tool "NO" Rules List

The adoption and use of an ATPG tool may automate and reduce the cycle time involved with the test pattern generation; however, ATPG tools also apply some restrictions on the design. These restrictions are commonly referred to as "design rules" or "test rules." The design rules related to ATPG fall into several categories: 1) rules required to allow the ATPG tool to operate; 2) rules that would enhance fault coverage or reduce ATPG tool runtime; and 3) rules that would result in "safe" vectors that do not cause driven contention or other destructive behavior.

In many cases, the specific set of design "rules" applied to a circuit depends on the specific tool being used (see Figure 2-17). Most tools require that the same universal set of restrictions be applied; however, some tools have developed "exception" handling for some of the common problems that "thwart" the ATPG process.

Here is an example list of "requirements" that would allow an ATPG tool to operate successfully on the design description to achieve high fault coverage with safe vectors:

No combinational feedback Combinational feedback creates many problems. It may cause the tracing algorithms or the fault simulation to enter an endless processing loop, the circuit may become an oscillator, or the circuit may exhibit "latching" behavior. For combinational-only tools, the latching behavior is a sequential element that is not handled by the algorithms, and the oscillator may represent a source clock signal that changes the internal state aperiodically to the tester input defined by the ATPG tool.

No non-gate-level logic Most ATPG tools base their analysis on the "gate-level" elements. Allowing transistor or analog models in the design description slows or hampers ATPG processing, and produces areas of the design description where a loss of controllability and observability occurs.

No non-deterministic logic Areas in the design description that support pulse generators, the X logic value, or the Z logic value interfere with the ability for ATPG to generate deterministic vectors that can achieve a measurable fault coverage for those areas. These areas also limit the ability to generate vectors for other logic that must be controlled or observed by the non-deterministic logic.

No clock signals used as data Using a repetitive periodic signal as a data signal in the combinational logic results in a pulse generator in that it can change the circuit's logic state at both the rising edge and the falling edge of the clock signal. A combinational ATPG tool generates a vector based on the application of a state (usually at the rising edge of some clock signal), and it expects that state to remain stable until the observe event (usually the next rising edge). The changing of the circuit state, independent of the ATPG tool, results in a vector that was created under bad assumptions.

No free-running internal clocks This rule is the precise statement of several of the other rules about non-determinism. Most ATPG tools do not have any way to model and evaluate a free-running clock.

No use of both edges of a clock This rule is also known as the "no negedge" rule. Purely combinational ATPG is accomplished independent of any clocking. However, there is an implicit assumption that the tester will synchronize the vector application based on a clock. Any use of both edges of the clock may result in the flawed assumption that logic states applied at the rising edge of some clock will remain stable throughout the cycle until the next rising edge of the same clock.

No dynamic logic Certain types of dynamic logic such as precharge-discharge and pullup-pulldown logic are viewed as self-timed circuits or contention-producing circuits by the ATPG engine. These types of circuits remove the determinism the ATPG tool bases its analysis on, or they compromise the safety factor in generated vectors. Dynamic logic circuits also limit the ability to conduct current-based testing such as iddq.

No asynchronous sequential logic Combinational-only ATPG, has no concept of sequential logic—sequential logic will be removed from the design description. For ATPG circuits that use a full-scan methodology (or some other methodology that renders the design description fully combinational), the sequential elements become control points and observe points. Using asynchronous set and reset sequential elements that allow the combinational logic to establish their state removes the implied stability on which the ATPG tool bases its analysis.

No non-transparent latches In some ATPG tools, latches are allowed. However, the latches must be purely combinational for the ATPG analysis, so they must be transparent during the vector generation and simulation processes. In most cases, the latch will effectively be replaced by a buffer in the ATPG's version of the design description.

No wire-OR or wire-AND logic Most ATPG tools will view "wired" logic as multiple source-driven nets that have the ability to create "driven contention."

No multiple-test-cycle timing paths The vectors created by the ATPG tool may or may not be created with the concept of timing, but the consideration of the timing of the circuit is important for the accuracy of the vectors. If the ATPG engine creates vectors that will be launched at the rising edge of some clock and if the result will be observed at the next rising edge, then combinational events that take longer than this test clock cycle to propagate will have erroneous results.

No gated clocks to sequential elements In many cases, the ATPG analysis of clocks in a sequential vector generator requires the clocking to be a periodic event. If the combinational logic can prevent a clock from occurring, then the ATPG tool may erroneously expect a sequential device to conduct a control or observe function.

Most of the rules that apply to ATPG also apply to DFT, testability, reliability, and the support of specific test architectures such as scan and built-in self-test. However, many more rules are specific to the other categories (these will be discussed in the scan, memory test, and embedded core test sections).

2.17.2 Exceptions to the Rules

The example set of rules described may seem somewhat restrictive. However, the rules exist for several reasons.

- The ATPG tools that are used to create vectors can't deal with certain logic constructs or timing (ATPG tools are generally zero delay or unit delay in their consideration of vectors—so timing information is not considered by the tools).
- The ATPG tools require deterministic circuit behavior within each defined clock cycle, so pulse generators, combinational feedback, cross-coupled gates, X-generators, and Z-generators may result in vectors that are not valid.
- Any structure that may create driven contention (a node being driven by two drivers to opposite logic values) can damage the final silicon during the testing process.

Some ATPG tools have created exception handling to some of the rules, or have modified the rules so that they are no longer "NO" statements. For example, some tools allow both the posedge and negedge consideration of clock signals by breaking the vector generation analysis into two time frames of one-half a cycle each. Some common exceptions are listed below:

Asynchronous set and reset elements Are allowed if the control signal sources directly from a package pin, or if a "test mode" signal is factored into the control signal that disables the asynchronous function during the application of ATPG vectors.

Gated clocks Are allowed if the control signal sources directly from a package pin, or if a "test mode" signal is factored into the control signal that disables the clock inhibit function during the application of ATPG vectors.

Transistor constructs Are allowed as long as they are simple and use a "unidirectional" transistor-primitive model, or as long as the structures can be modeled with tristate driver type logic element (data in, data out, and an enable signal).

Most rule exceptions are based on managing the problem areas in the design description, or by having the ATPG tool add algorithmic solutions to "handle" the problems.

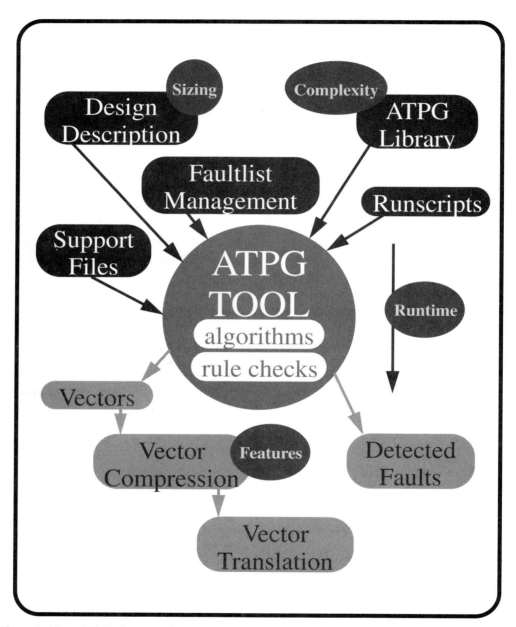

Figure 2-18 ATPG Measurables

2.18 Selecting an ATPG Tool

2.18.1 The Measurables

Two of the most difficult steps in adopting or supporting an ATPG methodology are 1) selecting ATPG tools for building a methodology, or 2) figuring out that your current methodology no longer meets your organizations needs and selecting new ATPG tools to replace existing ones. In both of these cases, the implication is that a "measurement criteria" can be used to determine "good" tools from "bad" tools. In reality, there exists no "good" or "bad," either a tool meets your needs, or it doesn't. The process that must be used for tool selection and ongoing review is one where the "measurables" are determined and assessed (see Figure 2-18). A mistake many organizations make is to measure tools against each other and then choose the best one—then this best tool does not fit their business case.

A more successful selection process is to determine the organization's business case, and from this determine the type of ATPG methodology needed. For example, if the business case is to make small application specific integrated circuits (ASICs), and all ASICs will be fully scanned, then the need is for a simple combinational-only vector generation tool, not for a full-featured, "can handle anything," sequential tool. The sequential tool, in this case, may require more work to set up for the design, may take longer to generate vectors, and may make more vectors than a simple combinational vector generator (or the vectors created may not be compressible).

In any case, the selection or review process is based on "measurables" involved with the ATPG tools. Some common measurables are listed below:

Netlist sizing How large of a design description can the ATPG tool accept and operate against? This metric is one of the most fundamental that must be assessed during any selection process, since it can limit the size and type of designs the organization can handle.

Memory image this metric is related to the Netlist Sizing metric in that it is the amount of computer memory that is required for any given design description. Even if the tool can accept a netlist larger than the target design, does the tool's internal representation of that design description exceed the normal computer memory available to the design or test organization? If the design descriptions overflow the memory available, then ATPG performance may be compromised since the computer must use "swap space" to conduct its analyses.

User complexity this metric is a relative metric in that there must be a standard of comparison. This category includes such elements as the number of files required to enable or conduct ATPG, the type of user interface (command line, graphical, script driven, batch system), and the available commands or functions.

Library support this metric is more about tool complexity. An ATPG tool must represent the design description in a "fundamental" format against which its algorithms are tailored to operate. The library may be measured by its richness (how many and what types of elements are supported), and the complexity involved with mapping any given silicon standard cell library to the ATPG library.

Features and fault models The heart of the ATPG engine is what algorithms it supports, and the value of the ATPG tool is what additional features and fault models are included within the tool. Some common fault models are: voltage—stuck-at, transition delay, path delay, bridging, and opens; current—quietness, toggle, and pseudo-stuck-at. Some common features included within ATPG tools are: algorithm selection (sequential versus combinational), faultlist management, rule checking, vector compression, vector diagnostics, vector addition (combining vector sets), schematic viewing, fault simulation, fault analysis, vector writing, and test point insertion.

Runtime Another fundamental measurement is, "How long does the ATPG tool take to create vectors?" This metric is sometimes separated into several categories:

- "How long does the tool take to process the design description?"
- "How long does it take to generate and simulate vectors?"
- "How long does it take to write out the generated vectors in the various output formats?"

Vector formats The number of vector data output formats, the available output options (scan serial, scan parallel, functional/sequential), and any vector translators available.

Vector sizing A feature available on some ATPG tools is the output data sizing features. Allowing the vectors to be written out in pieces and parts, to reduce the sizing of the individual patterns, eases disk requirements by allowing the vectors to be distributed across different disks.

Vector goodness Another critical metric is the goodness or accuracy of the vectors. If the ATPG engine is cycle-based, then negedge or self-timed events may not be understood or evaluated by the vector generator, and the vectors may mismatch when simulated or when applied on silicon. If the ATPG library is difficult to model, or if the library modeling language is not very comprehensive, then sometimes the library for ATPG may not fully match the library for other tools (such as the simulator)—ATPG vectors played back on the other tools will mismatch. The amount of work that must be done to "repair" a vector set after generation is also a valuable metric.

Performance against benchmark designs Many of the metrics can be evaluated in a stand-alone sense (for example, the tool can accept the size of your design, or it can't). However, many of the metrics are "relative" in nature and require a comparison against a known standard. For example, "What is a good runtime? Hours? Weeks?" Relative metrics require that some benchmark design descriptions should be used to provide a comparison standard. Several ATPG tools can operate on the benchmark circuits, and the results can be directly compared to each other.

2.18.2 The ATPG Benchmark Process

The well known and established process for selecting an ATPG tool has come to be called "benchmarking" (the conversion of a noun to a verb). Many tool vendors wish they could just state the merits of their tools, or provide statistics and metrics based on some industry-standard designs, and have the customer make a selection. However, each organization's business is different, and the selection of an ATPG tool should be based on the "real business." Therefore, the benchmarking process should be applied against "real designs."

Benchmarking is a very important part of the selection process, and every effort should be made to select the right tool for the business and to verify that the tool selected can operate on the common and extreme cases of the business. Note that benchmarking is also valid when making a "make versus buy" decision, as well as just making a "which tool to buy" decision. The steps involved with conducting an ATPG tool benchmarking process are as follows:

Establish the goals or needs of the tool Establish the needs, wants, or goals of the ATPG tool before starting. It is impossible to make a decision on a tool if the metrics are not established first. The goals should be established based on the business (what size are common or extreme designs? are the designs to be full-scan or not? what type of CPU and memory sizing will be commonly available? what typical netlist-to-vectors time schedule is expected? what fault models are needed?).

Select the benchmark circuits Select a set of designs that represent the real business. Do not rely solely on small circuits that may highlight, or break, some individual features of the tools—especially if the contrived circuit is not something that is generally supported by the business. If a real design can be used, then this is the best benchmark circuit. Note that the lifetime of the tool must also be considered—will the tool be used on just one major design and then discarded, or is an ATPG methodology to be applied by the organization on all future designs until the tool is no longer effective? The benchmark design circuits selected for use must be chosen based on the intended application of the tool.

Compare a variety of tools "One" does not make a comparison. Talk to the tool vendors (or developers) to understand what features are offered by the various tools and select those that meet the business needs. Then select several tools and run them against each other with the same benchmark circuits being used by all the tools. Generally, the vendors assist in the benchmark process, but a considerable amount of work is involved in porting an ATPG library and establishing the initial files and computer environment for each tool.

Fully evaluate the vectors Do not operate the tool to the point of creating vectors and fault coverage numbers, and then make decisions. Make sure that the vectors are valid. Simulate them, with timing if applicable, and if the benchmark circuit was an existing chip, then apply the vectors to the real chip on the tester (for this reason, it is wise to include a manufactured chip in the set of benchmark circuits). One of the most difficult aspects of an ATPG methodology is not the adoption of the tool, but the integration of the vector flow (vector translation, vector verification, vector diagnostics).

Create a matrix of important measurables Place all the important measurables in a tabular form so that the various tools can be compared directly. The usual measurables are: fault models supported; fault coverage per benchmark circuit; runtime per benchmark circuit; disk/RAM usage per benchmark circuit; and number of vectors created per benchmark circuit. Some example comparison matrices are in Table 2-1 and Table 2-2.

Fault models supported:

Table 2-1 ATPG Tool Benchmark Comparison Matrix for Fault Models

Tool	Stuck	TDelay	PDelay	Iddq	Other	Compression
Company A	Y	Y	N	N	N	N
Company B	Y	Y	N	Y	N	Y
Company C	Y	Y	Y	N	Y	Y
Internal	Y	Y	Y	N	Y	N

Stuck-at Fault Coverage and Number of Vectors per Benchmark Circuit:

Table 2-2 ATPG Tool Benchmark Selection Matrix for Fault Coverage/Vectors

Tool	CKT A 80,000 Faults	CKT B 42,000 Faults	CKT C 16,000 Faults	CKT D 120,000 Faults	CKT E 245,000 Faults	Overall Cvg. Ave.% Total Vectors
Company A # Vectors	92.6% 327	87.4% 126	99.5% 45	86.3% 856	42.9% 1244	66.67% 2598
Company B # Vectors	92.6% 298	86.9% 113	99.5% 43	87.1% 844	46.4% 1196	68.52% 2494
Company C # Vectors	96.5% 156	87.2% 89	99.4% 16	85.3% 789	52% 997	71.47% 2047
Internal # Vectors	98.9% 525	85.3% 212	99.3% 62	82.4% 907	62.4% 1376	76.06% 3082

2.19 ATPG Fundamentals Summary

2.19.1 Establishing an ATPG Methodology

Establishing an ATPG methodology is not just the selection of an ATPG tool, but it is also the development of a "concurrent engineering" process, where design rules and tools are applied at the front-end, middle, and back-end of a design process. The significant change involved with adopting an ATPG methodology is that it moves vector generation from its traditional place as a "back-end" or "post-silicon" process to a concurrent engineering "during-design" process.

The key reason to adopt an ATPG process is "economics." A concurrent engineering ATPG process can result in having the manufacturing test vectors and "qualification" vectors available before silicon arrives back from the manufacturing plant, thereby reducing the "Time-to-Market" or "Time-to-Volume" cycle time. The vectors delivered from an ATPG tool are also more "Cost-of-Test" efficient in that they are targeted at the design's structure, and if supported by a proper test architecture, they have more "Fault Coverage per Clock Cycle." More "Fault Efficient" vectors result in a smaller memory image in the tester.

History has proven that adopting an automated ASIC type design flow has reduced design cycle time, and has significantly reduced design cycle time of derivative products from an initial design. The ATPG process is part of this design automation philosophy and allows test to be a manageable and measurable "design budget" along with the traditional "frequency," "area/size," and "power." Items such as test time, number of test vectors, fault coverage, and the hardware impact of test can be tracked and measured during the design cycle by operating the ATPG tools on the prototype design during the development stages of the silicon device.

The heart of an ATPG process is the ATPG engine. The strength of the ATPG engine rests in its algorithms for tracing through the design description and establishing logic values. The ATPG engine relies on the fact that the design description does not stray from the bounds of a "process-able" format. To ensure that the design description stays within the bounds, "design rules" must be applied. Rule exceptions, extra algorithms, and extra processing steps such as testability analysis, fault equivalence analysis, and circuit learning can enhance the ATPG's engine in achieving higher fault coverage, fewer vectors, and a faster runtime.

To have a complete ATPG solution, several other related processes must be supported. An ATPG library must be developed, and this library must match the other simulation libraries for exact behavior and must be of similar structure to the gate-level elements used in the silicon library (to map faults from the design description to exact locations on the silicon). And a path must exist from the ATPG tool's vector format to the vector format for the other simulation environments and to the tester.

Finally, the adoption and selection of an ATPG solution requires the evaluation of the business model or business goals to select, or create, an ATPG tool and process that meets the needs of the business. Many organizations make the mistake of selecting the tool that does the most—or can handle the most complex circuits. Such tools carry a lot of internal overhead and may result in a slower processing time and a larger memory image, even for small designs. An ATPG tool selection process should be approached from a "measurable" point of view, and the correct measurable criteria should be established before collecting the group of tools together to conduct the "benchmark comparison."

2.20 Recommended Reading

To learn more about the topics in this chapter, the following reading is recommended:

Abromovici, Miron, Melvin A. Breuer, Arthur D. Friedman. *Digital Systems Testing and Testable Design.* New York: Computer Science Press, 1990.

Dillinger, Thomas E. *VLSI Engineering.* Englewood Cliffs, NJ: Prentice Hall, 1988.

Gulati, Ravi K. and Charles F. Hawkins, eds. I_{DDQ} *Testing of VLSI Circuits—A Special Issue of Journal of Electronic Testing: Theory and Applications.* Norwell, MA: Kluwer Academic Press, 1995.

Mazumder, Pinaki, and Elizabeth M. Rudnick. *Genetic Algorithms for VLSI Design, Layout, and Test Automation.* Upper Saddle River, NJ: Prentice Hall, 1999.

CHAPTER 3

Scan Architectures and Techniques

About This Chapter

This chapter contains the Scan Test Architecture and Scan Test Techniques sections and has been designed to teach the basic fundamentals of structured scan design-for-test and the techniques necessary to meet some common test goals and budgets.

This chapter includes material on scan-based testing, scan test styles, scan elements, scan shift registers (scan chains), scan test sequencing, scan vector timing, scan rules, safe-shifting, contention-free sampling, scan insertion, scan and multiple clock domains, scan and on-chip clock generation (PLLs), at-speed scan, critical paths, critical path selection, and scan-based logic built-in self-test.

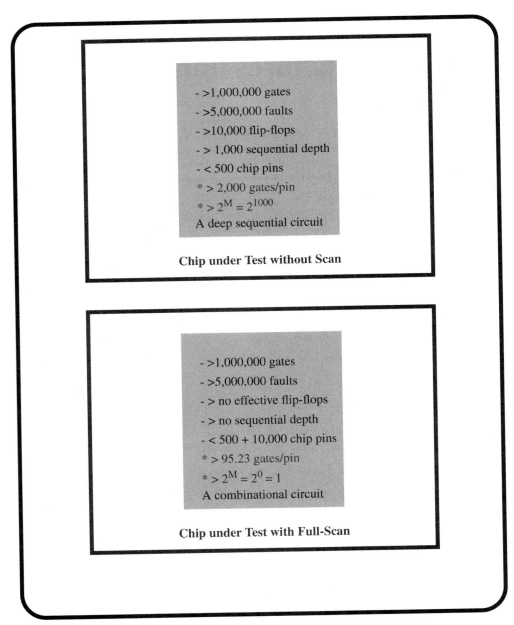

Figure 3-1   Introduction to Scan-Based Testing

3 • Scan Architectures and Techniques

3.1 Introduction to Scan-Based Testing

3.1.1 Purpose

The Scan Test chapter describes the basic components of scan testing and the scan shift architecture, and then describes the test techniques and supportive architecture components that are commonly used in parts of the industry today. The scan test architectures described in this section allow an ATPG process to generate efficient, targeted-fault-coverage vectors. This information is provided to clarify the trade-offs involved with the various scan test processes, rules, and restrictions, so that the correct techniques and architectures can be applied to meet cost-of-test, time-to-market, and quality level goals.

Note: This chapter will introduce multiple scan styles such as Mux-D, LSSD, and Clocked-Scan, but will concentrate on the Mux-D flip-flop style of scan. Most of the scan trade-offs and techniques described for the Mux-D style of scan are directly applicable to the other scan styles.

3.1.2 The Testing Problem

A serious set of test problems faces the manufacturers of VLSI components. History has shown that VLSI components have a tendency to get larger in the sense of gate density, pinouts, and wafer sizes, and smaller in the sense of feature sizes, drawn lines, and metallization connections. This means that the VLSI chip—like the "system-on-a-chip," made today and the ULSI chip made tomorrow—is more sensitive to contamination, process and metallization errors, and at the same time contains more physical possibilities for faults (more gates and gate connections). The problems of "higher frequency," "higher pin-count," "higher levels of integration," and "higher complexity," drive up the cost of the test platforms required to test modern chips, creating a "crisis in test cost." In certain markets, the recurring test cost is 2, 3, and even 4 times the silicon and package cost together.

One way of measuring the problem is to look at a "figure of merit": the gate-to-pin or fault-per-pin ratios. VLSI chips today have ratios such as 2000 gates per pin or 12,000 faults per pin. This means that a large portion of the chip is virtually untestable within a reasonable (cost-effective) number of functional sequential patterns and that pattern generation (algorithmic or manual) will be a long drawn-out process.

The best way to provide more access to buried logic is to provide more pins. Even though some recently designed chips have 500+ pins, this number is not keeping up with the growth in gate densities (and more pins mean higher chip cost). If a problem exists at 100K-gates with 200 pins, then going to 500 pins but increasing the gate count to 500K is not any better (and embedding huge complex microprocessor cores is not making this problem any easier, either).

Testing chips with this kind of density problem, functionally, through a limited number of pins, would be extremely difficult even if the chips were completely combinational, but adding sequential depth adds a level of complexity that is exponential in its effect on the test process. The key concerns are the amount of time it takes to generate vectors to achieve a targeted quality level (fault coverage), the number of vectors required to achieve that quality level, the cost or complexity of the tester required to play back the vectors, and the ability to achieve that quality level at all. These concerns are even more pressing if the chip in question must go into a low-cost market.

The best solution, when faced with high gate count, cost sensitivity, or sequentially complex chips, is to provide some sort of test access (control and observation) of internal nodes that will support the ability to reduce the test time, the test data, and the complexity of the test platform; or to enhance the ability to achieve a high quality metric; or to support the ability to automate the test or vector generation process. Several methods can be applied but, historically, the most effective is full-scan.

3.1.3   Scan Testing

The most fundamental requirement to allow a test process to assess or measure a high quality level, to conduct frequency-based binning, and to allow cycle time reduction with processes such as automated test pattern generation (ATPG), is to establish a repeatable test methodology or strategy that will be applied during the design and build process. For long-term gains in efficiency, it is best if the test methodology or strategy will be repeatable and consistent across all devices and designs. The ultimate goal of applying this test methodology, during the design and implementation phase of a chip, is to ensure that there are "hardware hooks" in the final design that would allow a tester to apply vectors that can achieve a high-quality measurement. Scan-based testing is such a methodology (see Figure 3-1).

Scan testing is simply defined as the process of using a scan architecture to conduct testing. Scan is known as a structured methodology, because it can be standardized, is repeatable, and is easily automatable (both in insertion and in vector generation). A scan architecture allows a data state to be placed within the chip by using scan shift registers, and also allows the data state of a chip to be observed by using those same scan shift registers. A scan architecture also allows algorithmic software tools to verify the test design's correctness and to create or generate the necessary structural test vectors required to verify that the rest of the chip has passed "defect free" through the manufacturing process. Multiple scan shift registers, or scan chains, help to optimize the vector depth required by the tester. Overall, a scan testing methodology can enhance the ability to achieve a high-quality metric, reduce the cost-of-test (vector data sizing), reduce the time it takes to get the vectors to the tester—and therefore, to get the chip into volume production.

3.1.4   Scan Testing Misconceptions

There are several publicly held misconceptions about scan design:
- that scan can negatively impact the performance of a device
- that scan can negatively impact the silicon area of a device
- that scan requires special "deep memory" tester options to operate efficiently
- that scan can be used only to test the structure of a device, but not its AC or frequency performance
- that scan testing is slow (it must be—it's serialized)
- that building scan into a device is a complicated process

If done incorrectly, a scan architecture, and scan testing using this "bad" scan architecture, can support these misconceptions. However, if the trade-offs involved with creating a scan architecture are properly considered and the proper optimizations are applied, then all of these "scan myths" can be debunked.

For example, some scan elements can be used that do not impact the setup-time or the clock-to-Q time of a register transfer—but these cells may have a size impact. There are also development methods that can be used to substitute scan sequential elements for functional sequential elements, and to route the scan control signals efficiently so that the area impact is significantly less than 5% (not the 18% that is commonly bantered around). These same optimizations allow scan data to be applied to the chip in an at-speed manner that both reduces test time and allows scan to be used to test AC fault models and specifications. Also, scan has become popularized in the industry to such an extent several tools are available to choose from to "insert" scan, and to produce vectors automatically using ATPG.

However, there are still those out there who can't separate their thought processes from "functional" test, and believe that scan is a way to serialize functional vectors—and that the development and application of these serialized functional vectors is a slow and painful process. This chapter will endeavor to show that scan is an optimized test architecture that is meant to be used for "structural" test and in conjunction with ATPG to create "high fault coverage" vectors in a short period of time; that techniques can be applied to reduce the vector data and test time, which ultimately reduces the cost-of-test; and that the installation of a scan architecture is not overly difficult, even given some complex logic structures and test rule violations.

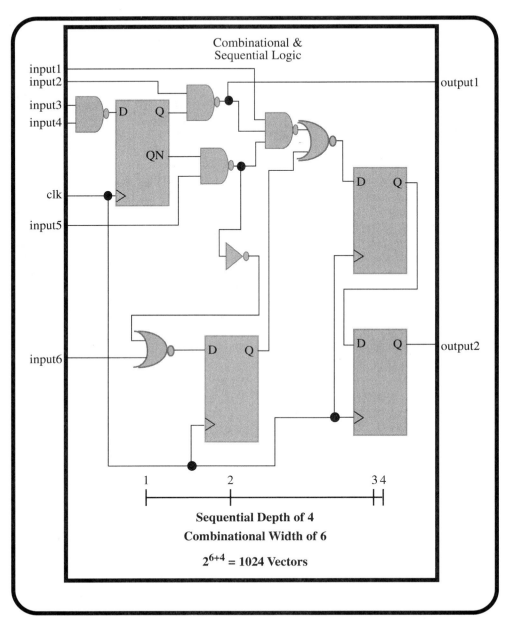

Figure 3-2 An Example Non-Scan Circuit

3.2 Functional Testing

Functional test is the application of functional, operational, or behavioral vectors to a circuit under test. If the vectors are not graded for structural coverage, then the vectors are generally referred to as "design verification" vectors. Many organizations still rely on functional vectors to accomplish final, manufacturing, or product test. These are fine if the device is small or if the majority of the functional vectors exist from a previous, but similar, design (or the functional vectors are easy to generate). However, there are several problems with the use of functional vectors for any kind of manufacturing test on modern VLSI designs.

First, functional vectors are very good for determining behavior, but are not especially good or efficient at structural verification—and the purpose of a manufacturing test should be to verify that the manufacturing process did not add any destructive defect content. The behavioral verification should be accomplished before incurring the expense of a mask set—not after the silicon is manufactured.

Second, functional vectors must be graded to verify the structural fault coverage. This is an extra task, and sometimes a significantly extra amount of work that must be accomplished prior to the development of the manufacturing test program. If the creation and delivery of functional vectors is not fast and efficient, then the final manufacturing test program may not be ready for months after the silicon is delivered—this delay negatively affects the time-to-volume (TTV) and parts sitting a warehouse waiting on vectors get more expensive over time. In some cases, the fault coverage of functional vectors has never achieved the required quality level and the final test program verifies the part with "quality" exceptions (insufficient coverage).

Note that most functional fault coverage assessment done is for the DC (stuck-at) fault model. Currently, there are fault simulators for the DC fault classes, but not many commercially available simulators can adequately time grade a vector set for frequency assessment against timing critical paths.

Finally, functional vectors are designed to verify circuit behavior, and so they are not as efficient as "deterministic" structural vectors. In modern "system-on-a-chip" and VLSI designs, the test data volume is becoming a critical aspect of test cost. Conducting a vector reload because the vector volume exceeds the tester's on-board memory can add seconds to test time (and test cost is measured by "in-socket" time).

As an example, Figure 3-2 shows a circuit with both combinational inputs and sequential depth. A mixed circuit like this requires both combinational inputs and some sequential clocking. To fully exhaustively test the example circuit would require 2^6 combinations of combinational vectors and 2^4 sequential combinations of the 2^6 combinational vectors. So this circuit could require 2^{4+6} vectors to gain 100% confidence without fault grading. Note that adding more pins or adding more sequential depth will add exponentially to the testing problem (2^{4+6} could easily grow to 2^{5+7}—or 1024 vectors could grow to 4096).

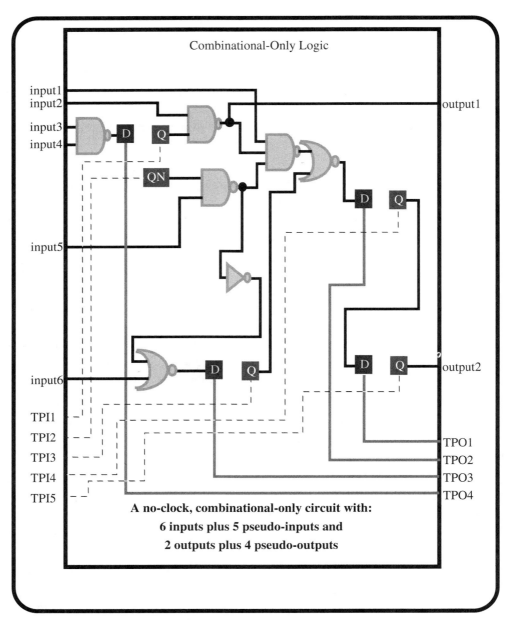

Figure 3-3 Scan Effective Circuit

3.3 The Scan Effective Circuit

The "scan effective circuit" is an optimization for most ATPG tools that converts the sequential design description into a combinational-only circuit. Scan gives direct access to internal nodes by converting every flip-flop to at least a pair of test points. The control test points are the Q outputs (and QBar outputs, if supported) of every scan flip-flop, and they allow circuit control values to be delivered directly (by conducting the scan shift-load operation). The observe test points are the D inputs of every scan flip-flop, and they allows circuit values to be observed directly (by conducting the scan shift-unload operation).

This "scan effective circuit" is a simplification for combinational ATPG tools (see Figure 3-3). ATPG tools can assume that the sequential elements are not even in the circuit, but that only control and observe points exist—the natural inputs and outputs and all other test points accessible by scan. In most cases, some other analysis, simulation, or rule checking will be done to ensure that the scan load-unload operations occur correctly, that the functional sample operation involving the now absent flip-flops is understood, and that hazardous or destructive circuit behavior does not result from the scan-shift operations (note: to optimize ATPG runtime, the load-unload may be checked once before vector generation, but it may not be further analyzed or simulated as part of the vector generation or vector delivery process of the ATPG tool).

The scan effective circuit is made by conducting the scan insertion process. The scan insertion process substitutes all standard system flip-flops with scannable equivalent flip-flops and connects them into scan chains. After a full scan design description is given to a combinational ATPG tool, the ATPG tool will convert the design description into an internal format that represents the scan effective (combinational-only) circuit.

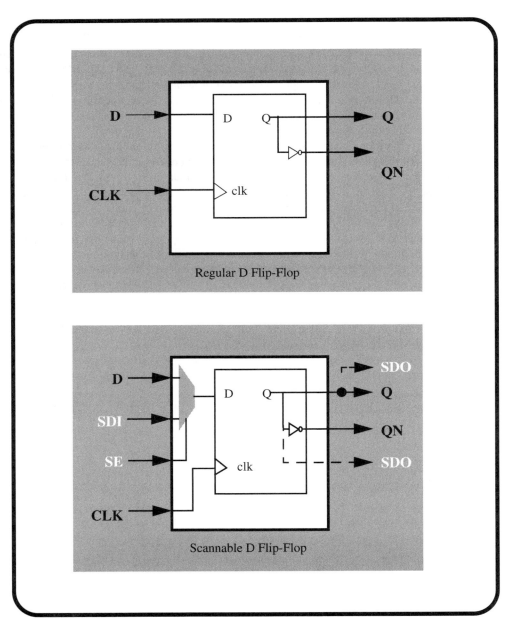

Figure 3-4 Flip-Flop versus Scan Flip-Flop

3 • Scan Architectures and Techniques   103

3.4 The Mux-D Style Scan Flip-Flops

3.4.1 The Multiplexed-D Flip-Flop Scan Cell

The Mux-D flip-flop scan cell is the most general-purpose scan cell used. It is basically just a flip-flop with a multiplexor in front of the D-input pin.

3.4.2 Perceived Silicon Impact of the Mux-D Scan Flip-Flop

Note that the Mux-D scan cell has an added SDI, SE, and SDO port and the SDO port may be dedicated or taken from the Q or QN ports (see Figure 3-4). This allows data to be scanned in, data to be scanned out, and the "scan shift/functional sample" mode to be selected. The perception is that the multiplexor will slow the flip-flop down (increase the setup time requirement), that the SDO connection may load down the output and slow the clock-to-Q or output data propagation time, and that the SDI, SDO, and SE connections will complicate the routing of the chip. These perceptions are true only if no optimizations are accomplished. The flip-flops should be made so that the input multiplexor is not unduly slow, and the routing of SDI, SDO, and SE can be inserted optimally to minimize the timing and wire route impact.

Also note that the impact of scan decreases as the feature geometry shrinks. Recently, the clock-to-Q speed of silicon library flip-flops has exceeded the routing maximum skew rating associated with clock-signal fanout-tree routing. In this case, the multiplexor in front of the D port on the flip-flop helps manage hold-time problems.

3.4.3 Other Types of Scan Flip-Flops

The Mux-D scan flip-flop is not the only scan cell that can be used for scan testing. It is the most widely accepted scan style, since almost all commercially available DFT and ATPG tools support the scan insertion, rule-checking, and vector generation for Mux-D flip-flop based scan (while having only limited support for other types of scan styles). Other scan cells can be used instead of the Mux-D Flip-Flop, and these cells have their different engineering trade-offs. Two of the most common cells are the Clocked-Scan cell and the LSSD Scan elements.

3.4.3.1 The Clocked-Scan Cell

A version of the Clocked-Scan cell incorporates an extra master latch within the D flip-flop to eliminate the multiplexer in front of the functional master latch in the Mux-D flip-flop type cell. This cell effectively places the scan multiplexer in front of the slave and adds a test clock to operate the extra master, and sometimes adds an additional signal to take over the functional slave. Since this cell is a three-latch design, it adds more cell area, and the extra test signals require extra routing. The positive aspect of this cell is that it does not add setup time to the functional path entering the D port of the flip-flop. The trade-off for this cell depends on the feature geometry—above .5 micron, the extra latch cell area has a larger impact than the clock routing impact, but may make a faster cell than the Mux-D flip-flop. Below .5 micron, the cell area impact is negligible, and the extra clock routing and clock route timing are more critical.

3.4.3.2   LSSD

The LSSD system is very similar to the Clocked-Scan system, except that the elements are loose latches. These cells can be used in latch-based or level-sensitive designs. LSSD is the acronym for Level-Sensitive Scan Design.

The LSSD system places master latches with a master clock separately from the slave latches and the slave clock. Different types of LSSD are supported in modern designs; some of them contain an extra scan-latch, with an additional test clock, and some of them use the functional master and functional slave, but with an additional test clock to operate the master. The masters and slaves (or scan-latches) are used to make the scan chain architecture, and test data launched from the system slaves is captured by the system masters for fault coverage assessment. The same trade-offs as for Clocked-Scan apply but with the addition of placing and routing separated master and slave latches with separate clock signals. LSSD should be used when a latch-based design style is adopted.

3.4.4   Mixing Scan Styles

Note that the mixing of different scan styles in the same design is an extreme complication that is not recommended. The "tool gymnastics" that must be done to accommodate mixing LSSD and Mux-D, or Clocked-Scan and Mux-D, do more to complicate the test flow than can sometimes be gained by using an ATPG methodology (unless the mixing is in isolated test areas, in which case, some other concerns become evident—mostly, how the logic between the different isolated scan domains is tested).

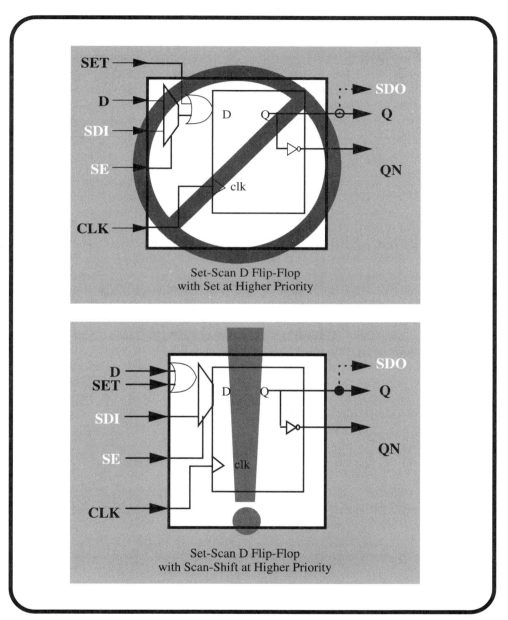

Figure 3-5 Example Set-Scan Flip-Flops

3.5 Preferred Mux-D Scan Flip-Flops

3.5.1 Operation Priority of the Multiplexed-D Flip-Flop Scan Cell

The Mux-D flip-flop scan cell is basically just a flip-flop with a multiplexor in front of the D-input pin. However, this cell cannot be made just by placing a multiplexor in front of the D input pin of any flip-flop. The Mux-D flip-flop should be made as a single unit because the scan operation must be the most fundamental (highest priority) operation of the flip-flop. If the scan operation is not the most fundamental operation, then external control signals and other additional safety logic must be applied to ensure that the flip-flop does not react to the system values applied during the scan shifting process.

For example, if a flip-flop within a scan chain has a set, reset, or hold function that has higher priority than the scan multiplexor, then as seemingly random data is shifted through the flip-flop (and other system flip-flops), the combinational logic of the overall circuit will react to the random state, and the flip-flop control signals may react to applied system values (see Figure 3-5). The fear is that the flip-flop would randomly set, reset, or hold during the shift operation, corrupting the input scan data (the input stream is no longer the known stimulus required to conduct testing). Scan elements made to be immune to external stimulus during the shift operation are known as "safe shift" elements.

To ensure that shift data corruption does not happen during the shift operation, the flip-flop needs to be constructed with the scan multiplexor as the highest priority function (closest to the master latch). If some operations have higher priority than the scan shift and the flip-flops are to be constructed as "safe shift" elements, then these operations need to be internally disabled while the shift control signal, SE, is asserted.

If the flip-flops are made with no internal "safe shifting" protection, then external gates must be added to ensure that the set, reset, or hold functions do not assert while SE is asserted.

3.5.2 The Mux-D Flip-Flop Family

The Mux-D type of scan flip-flop can come in many flavors, and generally the family of flip-flops does support all features and operations: Synchronous-Set; Synchronous-Clear/Reset; Synchronous-Hold Enable; combinations of Set-Clear-Hold; and combinations of Asynchronous Set-Clear-Hold functions (with scan at a higher or blocking priority). Needing flip-flops in a design with many different features and operations is not a reason to disregard scan testing.

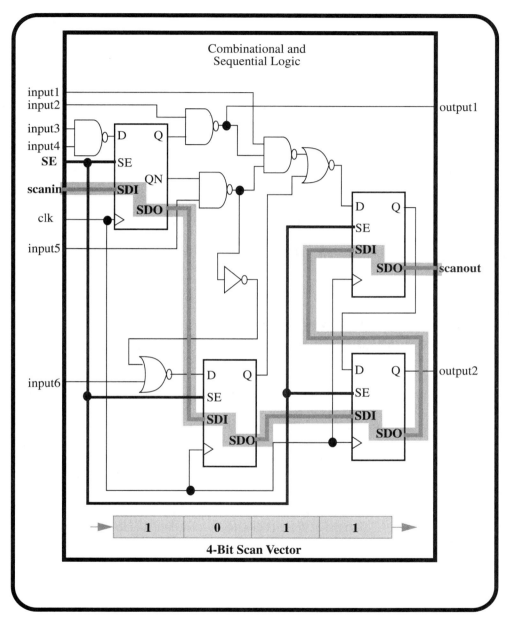

Figure 3-6 An Example Scan Circuit with a Scan Chain

3.6 The Scan Shift Register or Scan Chain

3.6.1 The Scan Architecture for Test

The basic premise behind the scan architecture is converting ordinary sequential elements within a design to scan sequential elements. There are several types of scan elements but all have one thing in common—they provide an alternate path into (SDI) and out of (SDO) the sequential element other than the normal parallel functional mode (D to Q). When all these alternate path signals are connected input-to-output (SDO to SDI), then a serial scan chain (shift register) can be constructed. If all sequential elements in a design are converted to scannable elements, then the test architecture is known as *full-scan*. If some non-scanned sequential elements are left in the design then the test architecture is known as *partial scan*.

Converting a functional design to a functional design with scan effectively converts each scannable sequential element into a primary input (control test point) and a primary output (observe test point). If all sequential elements within a design are converted to scannable sequential elements, then the design is effectively reduced to a combinational-only set of circuits surrounded by primary inputs and outputs. This simplification allows combinational ATPG tools to be used and to be more effective.

3.6.2 The Scan Shift Register (a.k.a The Scan Chain)

The scan shift register, or scan chain, is made by the connection of many scan elements into a serial shift register chain (see Figure 3-6). The scan chain is made by connecting scan cells together, beginning at a designated input pin, and tracing forward to the first SDI, and then traversing forward from SDO to SDI for however many flip-flops are defined in the scan chain—until finally the last SDO is connected to a designated output pin. The process of replacing regular flip-flops with scannable flip-flops and then connecting up the scan signals (SDI, SDO, SE) is known as scan substitution and scan insertion. Physical optimizations may be applied during scan insertion to make sure that the routing of SDI, SDO, and SE have minimal impact—for example, an algorithmic method may be used based on the closest physical location of one flip-flop to another (the "nearest neighbor" algorithm).

Note that the support of only one scan chain in a design is not an efficient test architecture. Although it allows ATPG, this type of design fuels the misconception of long test times and deep tester memory. A more economical optimization to a scan architecture is to use many scan chains designed to operate simultaneously to reduce the shift-depth, and therefore the tester cost (this allows normal parallel tester memory to be used instead of requiring a deep scan-only memory by spreading the vector data across many tester channels).

As can be seen in the example scan circuit in Figure 3-6, the state of the circuit can be fully defined by applying four clock signals to load a state (the scan vector) into the four scan cells in the scan chain—and on the fourth clock cycle, to also apply the necessary combinational inputs (then 1 clock cycle to conduct a sample). If two 2-bit scan chains were supported, then the test would require only two shift clock cycles plus a sample.

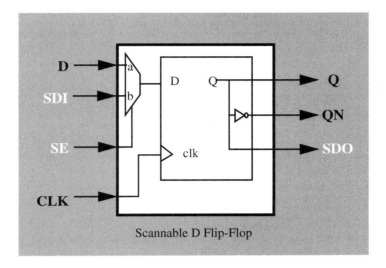

The scan cell provides observability and controllability of the signal path by conducting the four transfer functions of a scan element.

Operate: D to Q through port **a** of the input multiplexer: allows normal transparent operation of the element.

Scan Sample: D to SDO through port **a** of the input multiplexer: gives observability of logic that fans into the scan element.

Scan Load/Shift: SDI to SDO through the **b** port of the multiplexer: used to serially load/shift data into the scan chain while simultaneously unloading the last sample.

Scan Data Apply: SDI to Q through the **b** port of the multiplexer: allows the scan element to control the value of the output, thereby controlling the logic driven by Q.

Figure 3-7   Scan Element Operations

3.7 Scan Cell Operations

3.7.1 Scan Cell Transfer Functions

Scan testing is accomplished by changing the mode that a scan cell operates in to conduct one of the four transfer functions of a scan element. These four modes, shown in Figure 3-7, are: Functional Operation; Scan Sample; Scan Shift/Load; and Scan Data Apply.

Functional Operation The D-to-Q transfer through port **a** of the input multiplexer. This allows the normal sequential operation of the sequential element when the scan enable, SE, is de-asserted (normally defined as set to logic 0).

Scan Sample The D-to-SDO transfer through port **a** of the input multiplexer, where the scan enable, SE, is de-asserted to allow the D capture, and the next operation would be to assert SE to pass the data through SDO on the next clock cycle. This sequence gives observability of logic that fans into the scan element. Note that this is actually identical to the functional operation mode for some scan elements, except that scan shift operations occur before and after this mode.

Scan Shift Load/Unload The SDI-to-SDO transfer through the **b** port of the input multiplexer (the scan chain) when the scan enable, SE, is asserted (normally defined as set to logic 1). This operation is used to serially shift load/unload scan chain data through the scan cell.

Scan Data Apply The SDI-to-Q transfer through the **b** port of the input multiplexer when the scan enable, SE, is asserted for the SDI shift, and the next operation would be to de-assert SE to capture the result on the D pin of the scan chain collectively applying the scan state on all the Q pins. This operation allows the individual scan cell to control the value of Q and, thereby, controls the logic driven by Q.

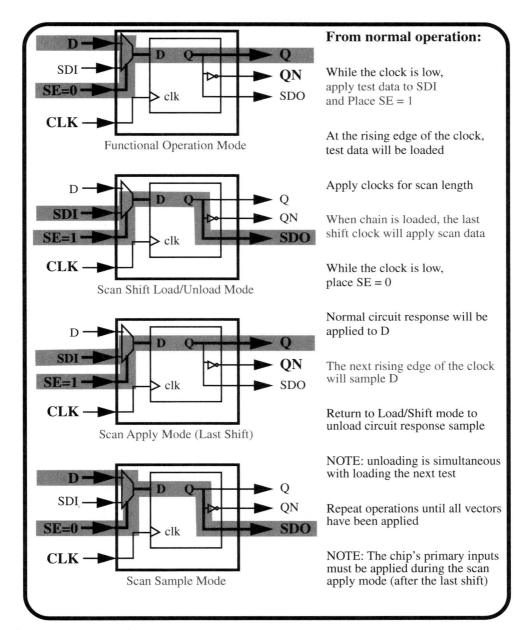

Figure 3-8 Example Scan Test Sequencing

3.8 Scan Test Sequencing

Scan testing is accomplished by applying the scan cell operation modes in a particular sequence (see Figure 3-8). A normal scan testing sequence is outlined below.

Functional Operation Initially the chip starts in functional operation mode. Regular chip data is applied to the parallel chip inputs and the clock signal toggles, which processes the applied pin input data sequentially through the chip.

Functional mode must be exited before scan testing can begin. Depending on the type of scan architecture supported, functional mode may be exited by configuring the chip to a specific chip test or scan test mode, or it can be exited by simply asserting the scan control signals (if they are dedicated signals).

Scan Load/Unload While the clock is low, the scan enable signal, SE, is asserted to choose the test data path. SE is a broadcast (global) signal that changes the data source on all scan elements in scan chains to which it is connected (there may be more than one SE signal, but any given scan chain must have the same SE signal connected to all elements on that scan chain).

Test data is applied to the pin that feeds the first SDI of the first scan element in the shift register. The chip is now in scan shift mode and (if it is a full-scan architecture) will no longer process valid parallel data sequentially to the outputs—random-seeming logic values will propagate throughout the chip and will be evident at the output pins not associated with the scan architecture. The logic under test is now rendered fully combinational.

The clock is toggled for the length of the scan chain. This action will serially shift data into the chip from scan element to scan element. Note that while new scan data is being shifted in, the existing data in the scan chain will be shifted out—the shifted-out data should be the values associated with a previous capture operation.

Also note that data may be applied to the inputs of the chip that are not involved with scan shifting during this time.

Scan Data Apply The last shift clock (the clock edge associated with shifting a logic value into the last scan element) will apply the system data to the circuit. During the scan shift load/unload operation, the parallel data pins not involved with scan shifting are held to a constant data value. At this point, the scan chain is applying a predetermined calculated data state directly to the internal logic, and the parallel pins are applying external chip test data from the tester. This is the point where the targeted fault is exercised.

Scan Sample When the test data is applied, right after the last shift clock edge, the scan enable signal, SE, is de-asserted to select the functional data path through the scan multiplexors. The functional data paths feeding the D inputs to the scan flip-flops should have the logic values associated with the circuit's natural response to the applied data state. On the next rising edge of the clock this data, including any fault effect, is captured into scan flip-flops. Note that if the targeted (or serendipitous) fault effects are driven to the chip output pins, then they are sampled at a strobe point, identified as a tester edge set, at a time prior to the capture clock (or after the capture clock if the clock-to-out is greater than a test clock cycle).

Return to Scan Load/Unload   After sampling the circuit response data, immediately after the capture clock edge, the scan enable is asserted again to select the scan path through the scan multiplexors, and the next test is serially loaded. Note that the loading in of new scan data unloads the captured scan sample data. The operations of Scan Load/Unload, Scan Data Apply, and Scan Sample are repeated until all scan vectors in a pattern set are applied (and all scan pattern sets are applied).

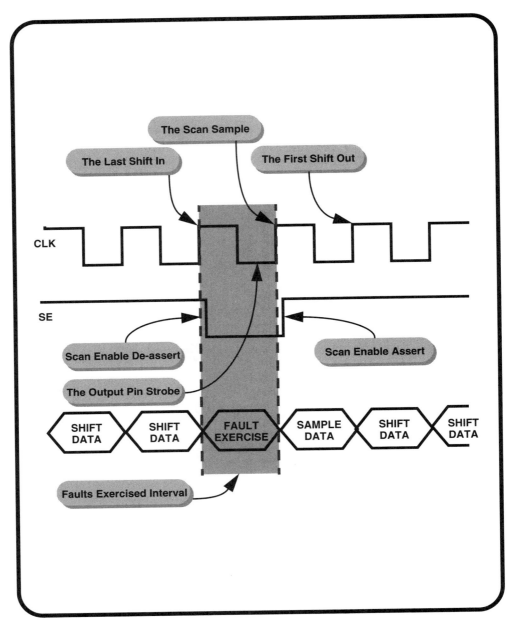

Figure 3-9 Example Scan Testing Timing

3 • Scan Architectures and Techniques

3.9 Scan Test Timing

The scan testing sequence can also be viewed as a timing diagram (note that Figure 3-9 is an idealized diagram, without setup and hold time consideration, to illustrate the various components of scan operation). The scan operations can be described in terms of a clock, scan data, and the scan enable signal(s). The key points in the timing diagram are as follows:

Scan Shift Data At each rising clock edge, new data is placed into the first scan element in the scan chain. This data clocked into the first scan element pushes all the data in the scan chain forward by one scan cell. The data being pushed forward exits the chip by passing through the last scan element in the scan chain.

Last Shift In The clock edge related to the loading of new data into the last scan element (or bit) in the scan chain (putting new data into the scan chains of exactly the same bit length as the scan chain).

Scan Enable De-assertion The de-assertion of scan enable after the last shift clock edge. This enables all scan flip-flops in a scan chain to sample the functional input port on the next valid clock edge (the scan sample clock).

Output Pin Strobe The point after the last shift clock edge that the output pins are strobed for valid data. This is important when fault effects are driven directly to output pins during the "fault exercise" interval.

Scan Sample The first valid clock edge after the last shift clock edge where system flip-flops will sample functional data because the scan enable is de-asserted. In other words, the clock edge where exercised fault effects are observed (captured) by scan flip-flops.

Scan Sample Data The data contained in the scan chain after the functional sample. This data is shifted out of the scan chain by the shifting in of new scan chain data. This is the "observe" data that must be evaluated to verify whether fault effects have been captured.

Fault Exercise Interval The interval between the last shift clock edge and the sample clock edge. The interval where the targeted fault models are exercised.

Scan Enable Assertion The assertion of scan enable after the scan sample clock edge, to allow scan flip-flops to begin shifting captured observe data on the next valid clock edge (the first shift out clock edge).

First Shift Out The clock edge related to the shifting out of the last scan element in the scan chain after the functional sample operation has occurred. Note that the first shift out of the sampled scan data coincides with the first shift in of the next test.

As the timing diagram shows, a scan vector is shifted into a scan chain as long as SE is asserted. Once the scan vector is shifted in (a number of scan clocks are applied that is equal to the number of bits in the scan chain), then the last shift clock (the last applied clock while SE is asserted) signals the onset of the "faults exercised" interval.

If the faults that are exercised are to be viewed directly at the output pins, then they will be strobed (observed) at the defined tester strobe time (which is usually prior to the scan sample clock). If the faults that are exercised are to be captured by scan flip-flops, then the clock that occurs while SE is de-asserted will capture the circuit's response to the scanned-in state. Once the sample data is captured into the scan flip-flops, the SE is again asserted, and each clock cycle of the next test being scanned in pushes the sampled data state out.

Scan can be likened to marbles in a tube—a number of marbles are pushed into the tube to fill the tube exactly—the marbles are all modified somehow when the tube is exactly full, and then a new set of marbles, which fills the tube exactly, pushes the modified marbles out for observation. Each marble represents one scan bit. A marble representing an erroneous change will be observed when it exits the tube at some time later, after the sample operation (the fail indication is not observed at the time of the scan sample operation, but later when the fail indication is shifted out of a package pin).

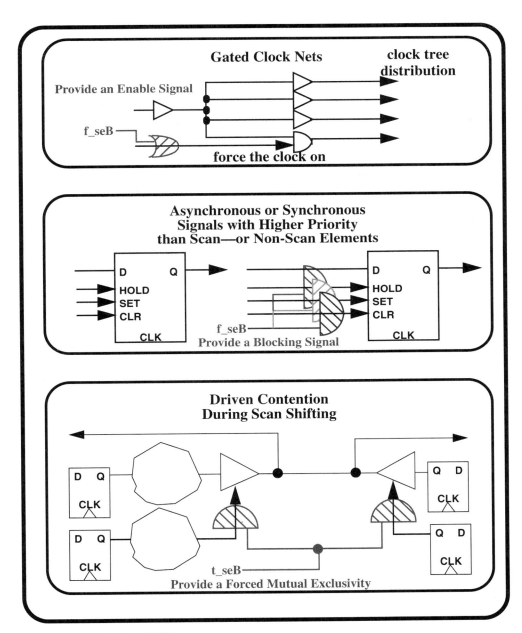

Figure 3-10 Safe Scan Shifting

3.10 Safe Scan Shifting

One of the simplifying assumptions of ATPG is that the shift operation will occur without problems—so the ATPG vector generation engine does not have to consider this part of the operation during its calculations. However, most ATPG rule checkers do have static checks for the shift operation. The reason is that some architectural features do exist that will prevent the scan shift operation from working or that may cause hazardous circuit operation. The ability to safely conduct a correct shift operation without hazardous behavior is known as "safe shifting" (see Figure 3-10). Safe shifting rule checks fall into the following categories: stable clocking of scan flip-flops, stable data to/from scan flip-flops, legal scan constructs, and safe circuit behavior.

Stable Clocking Since all scan flip-flops in the design must receive a clock signal directly from the tester to synchronize test operations and since scan flip-flops operate by clocking data from one scan cell in a scan chain to the next, a stable clock to all scan flip-flops is required. If the clock tree, or individual flip-flops, has any sort of clock gating that is not controllable during scan shift mode, then the ability to shift data through the scan chains may not be possible. There should either be: 1) no clock gating; or 2) a signal that is directly controllable by the tester that will disable clock gating (e.g., force_SE).

Stable Data Scan flip-flops conduct the scan shift operation by passing valid data from one scan element to another on each shift clock edge. During the shift process, the combinational logic of the circuit reacts to the random data state applied with each shift clock. If sequential elements in the design will react to the random applied state, then the scan chain will reset itself or have a blockage to passing scan data during the shift process. This result is especially true for sequential cells with asynchronous operations. A direct disable signal should be provided (e.g., force_SE).

Legal Scan Constructs Another data-handling problem that may exist is any "non-initialized" sequential logic. During the shift operation, any non-scanned sequential elements in the design may be receiving random input values from the combinational logic. Since the scan shift operation is not simulated or calculated during ATPG, these non-scan flip-flops (or similar constructs) will have a random or logic X state at the end of scan shifting. This violates one of the definitions of test, "a known state." These types of circuit features should not be supported, or if they are required, then they should be "set or reset"-able directly from the tester—a direct control signal should be provided (e.g., force_SE).

Safe Shifting Another key principle that must be adhered to during the shift process is that the shift operations should not damage the chip. The most common violation here is driving contention on multiple driver busses. During shift, only one driver should be enabled and all others should be disabled (e.g., tristate_SE to force mutual exclusivity).

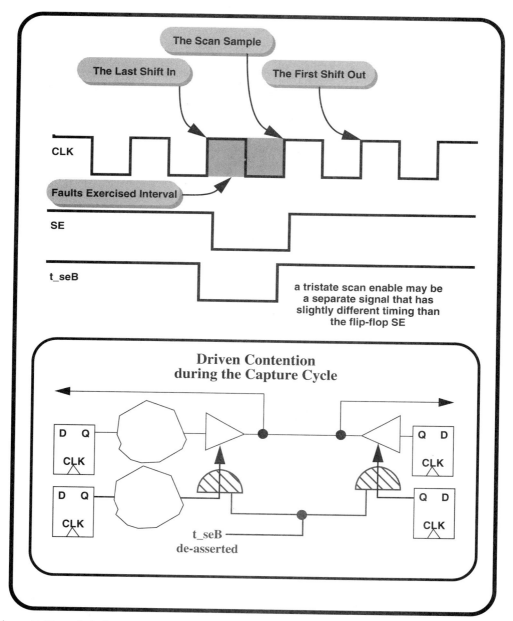

Figure 3-11 Safe Scan Vectors

3.11 Safe Scan Sampling: Contention-Free Vectors

3.11.1 Contention-Free Vectors

Although care must be taken to prevent driving contention during the shift mode (and this must be done with test architecture elements such as providing a forced mutual exclusivity signal—the t_seB signal in Figure 3-11), it is the ATPG engine that must make the decisions during the scan sample interval. In an ideal sense, when the scan enable is de-asserted, the circuit should resolve to a completely functional architecture. If the ATPG solution to exercising a fault, and then the propagation of that fault effect to an observe point, requires the ATPG tool to make a vector that drives contention during the "exercise the fault" or "the sample" interval (whenever the SE or tristate_SE signal is de-asserted), then this vector will be created by the ATPG engine. If the vector is not checked for contention, then the ATPG tool will deliver unsafe and destructive vectors. For this reason, most ATPG tools will check for driven contention during the vector fault simulation mode that it uses to drop "serendipitous" fault coverage from the faultlist. Contention will cause the vector generator to discard vectors and the fault coverage associated with them, and these faults will be added back to the faultlist.

Note that some ATPG engines consider the existence or propagation of an "X" logic value as a contention condition. This has ramifications to successful vector generation. For example, if the circuit model elements do not fully line up with the silicon behavior (X mapping is not identical), then contention conditions may be more prevalent than they would be in reality and the ATPG engine may be discarding perfectly good vectors. This result may even lead to a loss of coverage condition (no amount of vector generation will make up for discarded coverage due to contention checking). Also note that excessive contention identified by contention checking can also affect the runtime performance of the ATPG engine.

If a static "safe shift" architecture is supported that does not allow the tristate elements to ever resolve to a functional state, then the sample cycle may be safe, but fault coverage is lost by blocking controllability or observability to certain parts of the circuit.

3.11.1.1 Safe Sampling

Of the described scan rules, the two to be concerned with during the sample cycle are the "uninitialized state" (a logic X from a non-scanned sequential element) and the "no driven contention" rules. Driven contention is hazardous to the circuit, and the uninitialized state causes ambiguity and may affect observability and controllability.

3.11.1.2 True Sampling

The forcing constraints applied to the other scan rules for safe shifting (asynchronous operations, operations with higher priority than scan shift operations, and clock gating) do not need to be carried over into the scan sample interval. In fact, for true sampling (measuring faults in a true functional environment), the scan control signals involved with safe shifting should be de-asserted. If a reset or gated clock is scheduled by the ATPG tool, then this should be allowed to occur—this would provide the structural fault coverage of these logic functions.

3 • Scan Architectures and Techniques

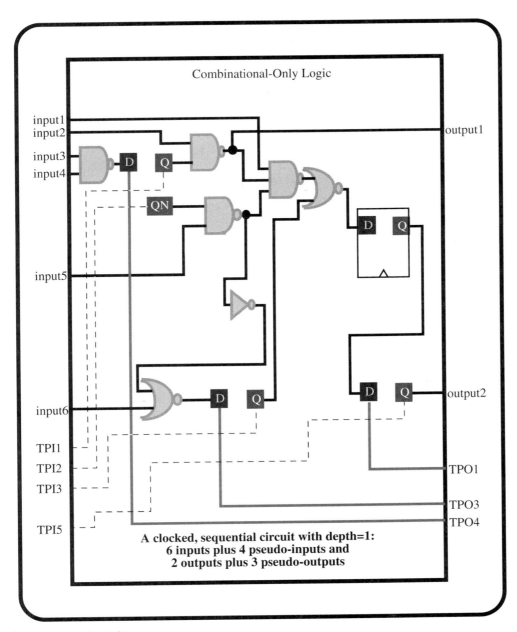

Figure 3-12 Partial Scan

3.12 Partial Scan

3.12.1 Scan Testing with Partial-Scan

As was defined earlier, scan testing is the process of using a scan architecture to conduct testing. Since the scan architecture is a serial shift register path made of all the scannable sequential registers in a design, the design description can be converted to a combinational-only representation for the ATPG tool. All the sequential elements within the circuit, since they are all involved with scan, can be loaded, shifted, and analyzed for correct operation and "safe shiftness" as a separate operation outside of the central ATPG process to support this combinational-only representation.

This efficiency step breaks down if there are non-scan sequential elements in the design description. Scan flip-flops are directly accessible for scan-in and scan-out, but the non-scan elements are not directly controllable or observable (except in certain conditions where they "shadow" a scan element by direct connection to the scan element). These non-scan elements leave areas of non-intialized circuitry after the shift process has occurred, resulting in loss of controllability, observability, and fault coverage. For example, during the scan sample interval, the elements left uninitialized after the scan shift process may cause a loss of controllability (an X in an element needed to exercise a fault or enable a propagation path) or may cause a loss of observability (the capture register is not scanned, and so is not observable). The controllability aspect can be relieved somewhat by applying a set or reset to the non-scan elements, but this usually allows only a known state of "one possible" logic value when entering the sample interval.

Any scan architecture that contains non-scan sequential elements (other than specialized memory circuitry) is known as a partial-scan architecture (see Figure 3-12). There are two main forms of partial-scan: the type where a netlist is analyzed and only a few cells are converted to scan cells—this is the original definition of partial scan; and the type where a netlist has all elements converted to scan except for a few in critical timing areas, or those involved with area impacts such as large register files—this is known as "mostly full-scan." The point to make here is that any non-scan elements left in the design description fed to the ATPG tool may cause loss of fault coverage, increased runtime, and an increase in the number of vectors created (non-scan sequential elements may thwart some forms of vector compression).

3.12.2 Sequential ATPG

There are many ways to approach the problem of loss of controllability and observability when non-scanned elements need to exist in the design. Many of these solutions involve manually "crawling through the netlist" and figuring out what kind of control and observe logic or ad hoc test points, need to be added to make an "exception" out of every non-scan instance (find ways to eliminate the effect of the non-scan element). This very slow process can't always be supported by modern design schedules. The more comprehensive solution is to move to a partial-scan or sequential ATPG tool.

To solve the uninitialized state problem adequately, the ATPG tool must become sequential; analysis and vector generation must occur on a netlist that includes sequential elements—the tool must now know how to put values into flip-flops using the natural set-reset-hold-update and clocking functions that flip-flops may support. For true analysis, some part of the scan shift operation should be simulated, or a full circuit state analysis must be done after the last shift event (to establish what the state of non-scan sequential elements may be after the shift operation). Adding these kinds of operations to the ATPG tools can significantly impact compute runtime (a few hours of full-scan analysis and ATPG can turn into days of sequential analysis and ATPG).

Sequential tools capable of handling partial-scan, and even no-scan, architectures are available commercially. However, if time-to-market and/or vector-data sizing is a concern, then partial scan is not an optimal solution because traditional partial-scan requires an amount of work to select which flip-flops to scan and which not to scan. Compute time and vector data also increase. In general (history has shown), if the design team selects which flip-flops not to scan, and this selection is based on the impact with design budgets, then the flip-flops chosen to scan are invariably the easy ones for ATPG, and the ones not chosen are the hard ones for sequential ATPG (leading to loss of coverage and increasing tool runtime).

There are cases, however, when devices are extremely area- or performance-constrained, and the cost-of-test is not as great of an issue. In these cases, partial-scan may be the best, or only, solution to achieve a high fault coverage under the applied constraints.

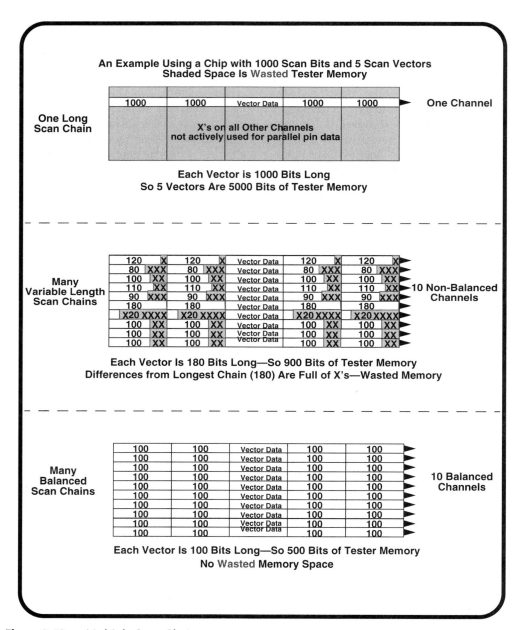

Figure 3-13 Multiple Scan Chains

3.13 Multiple Scan Chains

3.13.1 Advantages of Multiple Scan Chains

There is a trade-off between having one scan chain containing all the chip's sequential elements versus having several scan chains and dividing the number of flip-flops by the number of scan chains. The trade-off is the number of pins used to provide scan access versus test vector depth (and therefore, test time).

For example, a chip with 1,000 flip-flops and only one scan chain, would require 1,000 clock cycles to load in a scan state; however, using 10 scan chains would place 100 flip-flops in each scan chain and cost only 100 clock cycles to load in a chip state (if the scan chains are balanced to be the same length). The difference between the two architectures is that a single scan chain requires only 1 scan data input, 1 scan data output, and 1 scan enable, whereas 10 scan chains require 10 scan inputs, 10 scan outputs, and 1 scan enable (in this case, 3 pins versus 21 pins).

The cost savings in vector depth comes from distributing the vector data across multiple tester channels (the memory behind a tester driven input pin)—in Figure 3-13, 1 tester channel with 1000 bits of data per scan vector versus 10 tester channels with 100 bits of data per scan vector. A subtle fact that factors into this analysis is that for a given circuit, the number of vectors generated by ATPG is the same no matter if there is 1 scan chain or 100 scan chains—so it is more "cost-of-test" effective to distribute this data across more tester channels than to concentrate it in only 1 tester channel.

The trade-offs that factor into the support of one versus multiple are:

- Does the scan test architecture require dedicated package pins?
- Or will the scan test architecture borrow functional pins during a scan test mode?
- Is the targeted tester one that has a "deep scan memory option"?
- Or is the targeted tester one that supports normal depth functional memory?

If the product goals are to support a few dedicated test pins and the tester supports a deep memory option, then supporting fewer scan chains is the better solution. However, most designs target testers with limited functional memory, and conducting several vector reloads is very expensive in test time. So in most cases, it is more cost effective to support as many scan chains as possible and to borrow functional pins. It is still required to have at least one dedicated test pin to place the part in scan test mode (or to rely on an existing test type interface such as JTAG), but this is less expensive than the 3 pins required to have one scan chain with dedicated pins.

3.13.2 Balanced Scan Chains

When multiple scan chains are supported, they should be balanced. Balanced means that all scan chains contain the same number of scan bits (with the exception of one short scan chain for the odd number of leftover bits—designers hardly ever put just the right number of flip-flops in a part to divide evenly by the number of scan chains to be supported). Balancing scan chains is cost efficient in that it makes the best, most efficient use of tester memory.

When scan vectors reside in the tester, they are organized in parallel strings of scan loads (all the bits needed to load one scan chain). If all the strings are the same length, then the tester memory is fully used. If the strings are of various lengths, the tester must align them so that the last shifts all occur at the same time (the state is installed into the device all at one time). The alignment of the last shift requires reserving the tester memory to the longest scan chain's bit-length. This means that the shorter vectors must be "X-filled" or "X-padded" at the front (the part of the vector that enters the device at the beginning of a scan load). The Xs waste tester memory.

When an effort is made to make all scan chains the same bit-length, then the scan chains are said to be balanced. When all scan chains are balanced, then the tester memory does not contain any "X-filler."

It is best to divide the flip-flops into scan chains in such a way as to leave only one short scan chain if there are flip-flops left over. If the remainder scan chain is longer than the rest of the scan chains, then all those scan chains must be expanded with X-filler to meet the length of the odd remainder chain. It is better to have one short scan chain with X-filler than to X-fill all the other scan chains.

3 • Scan Architectures and Techniques

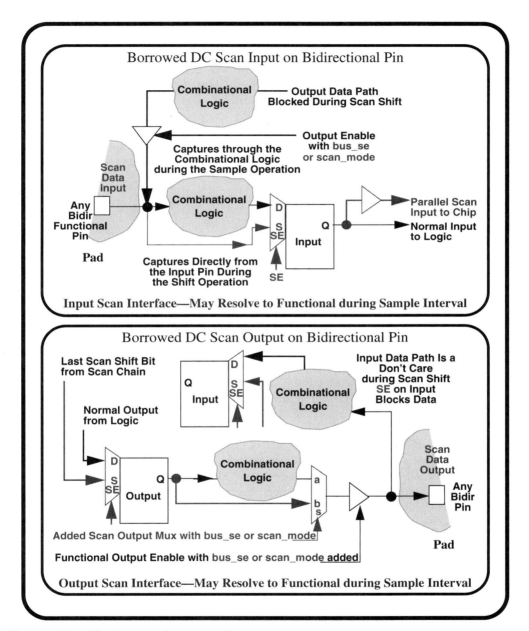

Figure 3-14   The Borrowed Scan Interface

3.14 The Borrowed Scan Interface

3.14.1 Setting up a Borrowed Scan Interface

If only one scan chain is used, the shift depth is long—and the test data depth and test time are negatively affected. However, only three pins are required for the scan interface. These pins are the single scan input port, scan-in, the single scan output port, scan-out, and the scan shift-sample control signal, scan enable.

If multiple scan chains are required to reduce the shift cost (number of clock cycles to load-unload a scan vector), then multiple scan input and scan output pins are required. In some cases, the adding of dedicated pins to a package for test purposes would require moving to a higher-cost (more pins) package. Another concern with dedicated scan test pins is that a set of pins will exist at the package interface that will not be used by the customer, and these pins must be labeled in the data book (the scan interface is generally only for manufacturing test, since scan can be used to "reverse engineer" the chip by register mapping).

A better, more cost effective solution is to "borrow" or "share" functional pins as the scan interface (see Figure 3-14). This requires starting the multiple scan chain inputs on certain functional pins, ending the multiple scan chain outputs on certain functional pins, and borrowing or dedicating a pin to implement the scan enable signal. This limits the number of scan chains to the number of available data pins supported by the chip's package.

There are two concerns with borrowing functional pins for the scan interface. One concern is mapping different types of pins to the various scan signals—for example, a scan input being shared with a functional input, output, or bidirectional pin. Another concern is that test coverage will be lost if the functional pins are used for scan purposes (the functional use of the functional pins will not be verified during test mode).

These concerns can be addressed in the DC scan environment (no timing considered) by using two methods: 1) specifying a scan input and scan output interface architecture that resolves to the functional state during the sample operation and that is held to a shift-only state during the shift operation; and 2) placing observe and control registers to recapture the lost coverage. The worst-case problem is to place a scan input interface connection or a scan output interface connection on a bidirectional functional pin. Solving this problem shows how to support any combination of scan inputs and outputs sharing any combination of functional pins, so this interface will be described. Note that this problem can be solved for the AC test environment as well, but this will be covered later in this chapter.

3.14.2 The Shared Scan Input Interface

For the case of the scan input pin being mapped to a functional bidirectional pin, the scan requirement is to constantly make sure that this pin is an input whenever the scan chains are shifting (SE is asserted). This requires that the functional output tristate driver must be held to high impedance during the shift operation so that the input can be used as a scan input—it can be released for the sample interval.

The scan insertion process should replace the natural input flip-flop with a scan flip-flop and should connect SDI to the input pin in front of the combinational logic (as close to the pad as possible). The global route scan enable, SE, connection should also be made. The tristate driver can be held with either a constant scan_mode signal that never de-asserts during all scan shifting or a dynamic signal can be used that mimics the SE signal to allow the interface to resolve to a functional state during the sample cycle.

If the constant scan_mode signal is used, then two scannable observe registers are needed to recover the fault coverage lost by blocking the effect of the output enable and the blocking of the output data. These registers have no functional purpose and should be placed at the data and output enable connections of the output tristate driver. Note that if the registers are placed in the hardware description, HDL or RTL, then they should be "don't touched" so that the synthesizer does not optimize them away since no functional Q connection exists—the Q connections will be connected to the scan architecture during the scan insertion process as SDO connections.

If the dynamic bus_SE type signal is supported, then this signal needs to be separate from the globally routed SE signal, because the timing is different between the assertion and de-assertion of internal scan multiplexors and the tristate drivers at the chip pin interface. Sometimes the difference is great enough that no "sweet spot" serves both architectures. The dynamic signal must be described to the ATPG tool to allow it to understand that it may convert a scan input to a functional output during the scan sample operation. This is not easily handled by all available tools.

3.14.3 The Shared Scan Output Interface

For the case of a scan output pin being mapped to a functional bidirectional pin, then the scan requirement is to ensure that this pin is always an output whenever the scan chains are shifting (SE is asserted). This condition requires that the functional output tristate driver must be held to the driving state all during the shift operation so the output can be used as a scan output—it can be released for the sample interval.

The scan insertion process should replace the natural output flip-flop with a scan flip-flop and add all scan connections. Scan insertion should also add an output multiplexor to bypass the output combinational logic. The tristate driver can be held in drive state with either a constant scan_mode signal that never de-asserts during all scan shifting or a dynamic signal that mimics the SE signal to allow the interface to resolve to a functional state during the sample cycle.

If the constant scan_mode signal is used, then two scannable observe registers are needed to recover the fault coverage lost by blocking the effect of the output enable and the blocking of the combinational logic after the last scan flip-flops. These registers have no functional purpose and should be placed at the functional data input to the scan test multiplexor and the output enable connection of the output tristate driver. Note that if the registers are placed in the hardware description, HDL or RTL, then they should be "don't touched" so that the synthesizer does not optimize them away since no functional Q connection exists—the Q connections will be connected to the scan architecture during the scan insertion process as SDO connections.

The same restrictions are to be applied if the dynamic bus_SE type signal is supported: the signal needs to be separate from the globally routed SE signal; and the dynamic signal must be described to the ATPG tool to allow it to understand that it may convert a scan output to a functional input during the scan sample operation.

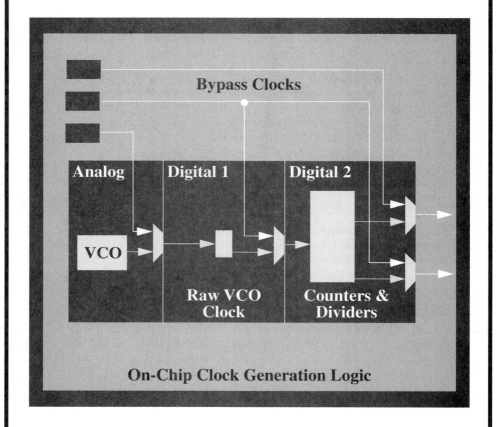

Figure 3-15 Clocking and Scan

3.15 Clocking, On-Chip Clock Sources, and Scan

3.15.1 On-Chip Clock Sources and Scan Testing

Two major problems occur with on-chip clock sources: basing testing on clock-out specifications, and any testing that requires starting and stopping the clock. In general, using an on-chip clock to conduct testing is a bad idea—especially for scan. Most economical test programs are based on the "input clock specification" method, where the test events (new vector data application and output vector strobe) are controlled and aligned to a synchronizing clock signal, input to the chip, by the tester. If the test is based on an internal clock, then the tester must conduct extensive edge searches to align the application and strobe points associated with each pin to a clock-out signal (basically, the tester must characterize the timing involved with the clock trees).

Another problem with any testing and the on-chip clock source has to do with tests that require clock starting and stopping. For example, the initial scan vectors cannot be applied until an on-chip PLL has output a "lock" signal informing the tester that it is stable; then at some point during the testing, if a clock-pause is required to test retention or Iddq, then the clock must be stopped or disabled. The problem comes in the re-starting of the clock—there may be unstable clocks, or several thousand clocks, before testing may proceed again. These extra or unstable clocks may disrupt the state of the machine, thereby invalidating the intention of the test (destroying the observe data).

To solve these problems, the most common solution is to require bypass clock inputs for scan testing (see Figure 3-15). This allows a tester to synchronize the new data application and the output data strobe to a clock signal generated by the tester. Note that the bypass test clock signals should allow the test clocks to be applied with the same skew and delay management as the original clock output(s) from the on-chip clock source.

There is one case where the on-chip test clock must be used for testing purposes, and that is if the clock frequency requirements of the device exceed the clock generation, clock signal edge rate, or data rate capabilities of the tester, and the device must be operated at its rated frequency during the test process (for example, AC verification is required). In this case, some data rate games can be played on the tester (such as assigning two tester channels to one package pin), and the device clocking may rely on the on-chip clock source.

3.15.2 On-Chip Clocks and Being Scan Tested

The ideal method to apply to an economical scan test architecture is to supply a bypass clock, or clocks, that allows the tester to drive all scan logic directly. The question then becomes "How is the on-chip clock source tested if it is being bypassed?" This is one of those situations where attempting to scan test all chip logic breaks down. In most cases, the clock generation logic is "black-boxed" or bypassed as far as scan testing is concerned.

However, the part of the on-chip clock generation logic that creates other clocks by digital logic division or multiplication can be scan tested, because these can get their clock source from the bypass clock. But the flip-flops that get their clock directly from an analog source such as a VCO cannot be scanned, because the VCO is not controllable during scan mode. These flip-flops can be scanned if the clock bypass is placed at the point where the VCO clock is input to the digital logic (however, if a bypass clock is input at this point, care must be taken to separate the digital power from the analog power if a scan-based current measurement technique such as Iddq is to be supported).

Generally, the on-chip clock generation logic may be fully bypassed and black-boxed, or it may be partially scan-inserted. Even though it may be partially scan-tested, it still may need its own test control modes and test logic to verify analog type testing involved with clock frequency, edge-placement, edge-rate, and jitter. This case is one exception to the ideal goal of having only scan vectors in the test program.

3 • Scan Architectures and Techniques 133

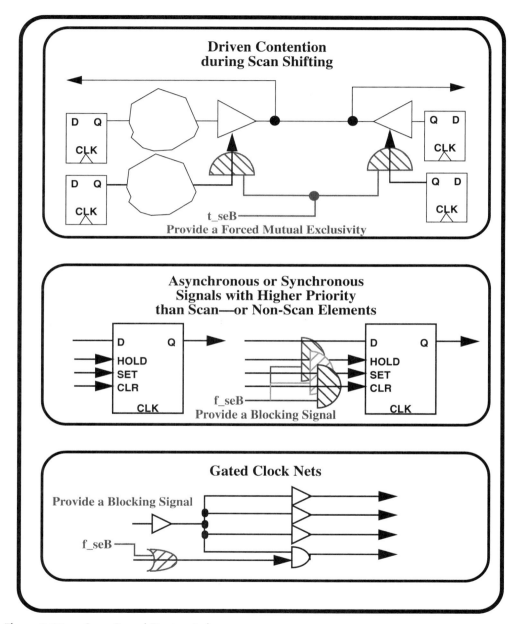

Figure 3-16 Scan-Based Design Rules

3.16 Scan-Based Design Rules

3.16.1 Scan-Based DFT and Design Rules

All the previous information concerning "safe-shifting," "contention-free scan sampling," scan interface architecture requirements, scan clock requirements, scan cell selection, and other requirements applied to successfully conduct scan testing, are generally collected together and placed in a document as a guide to a chip design group. This collection of requirements is called "the scan test design rules." Design rules must be applied for several reasons: to ensure that a scan architecture will operate correctly; to ensure that scan testing will not be destructive to the chip; but mostly because of limitations in either the tester or the ATPG tool.

When most engineers hear the term DFT, their thinking process is "the amount of work and type of work needed to implement the aforementioned design rules." In many cases, this is true. What used to be called design-for-test (DFT) is actually more accurately referred to as design-for-ATPG, because many of the current ATPG tools will handle a great many exceptions to what used to be "testability rules," so the job of DFT is to apply the design rules necessary to ensure that the ATPG tool can successfully operate on the design description and generate high fault coverage vectors.

Included below is a sample listing of a set of design rules that are applied because of ATPG tool limitations, tester limitations, and test cost budget restrictions (see also Figure 3-16). Test design rules are usually written as "NO"—you can't do that, so they have also been called, "the NO list." This illustrative set included below is overly restrictive to show a worst-case application where a limited tester or a weak ATPG tool may be establishing the test limitations. Also note that some of the rules are repeated— to show all of the areas of consideration that these rules violate.

3.16.2 The Rules

Posedge Synchronous Flip-Flop Scan Design Only (meaning)

- *Posedge*
    - No negedge clocks by inversion of a positive clock
    - No negedge by using negative sequential elements connected to a positive clock
- *Synchronous*
    - No asynchronous set, reset, or hold elements by using sequential elements with asynchronous functions or functions with higher priority than the scan operation
- *Flip-Flop*
    - No synchronous latches (gated with clock or with a combination of data and clock)
    - No transparent latches (gated with functionally derived signals (data or control)
    - No sequential elements made from cross-coupled gates

3 • Scan Architectures and Techniques 135

- No sequential elements made from combinational feedback
- No sequential elements represented as transistors
- **Scan Design**
  - No sequential elements that have no mapping to a scannable library element

Safe Scannability Design Style Only (meaning)
- *Test Mode Selection Stability*
  - No registered test mode logic (all scan and test mode logic is fully combinational from direct pin interface control)
  - No power-up reset type of test mode entrance or switching
  - No reset dependency during scan (the functional reset is just another signal or pin to be toggled during scan mode; the reset in no way hinders scan mode)
  - No test mode select logic that can be changed during scan operation by scan operation (scan-dependent mode logic)
- *Contention Free Vectors*
  - No tristate busses
  - No internal nodes that support logic values other than logic 1 or logic 0 (for example, no Z state and no chronic X generators)
  - No multiple driver nets
- *State Stability*
  - No combinational feedback
  - No non-scanned sequential devices (no partial-scan)
  - No self-timed logic
  - No multiple-cycle timing paths
  - No asynchronous pin-to-pin paths (e.g., input-to-output paths with no registration between)
  - No dynamic logic (pre-charge, pullups, etc.)
  - No non-constrained memory logic (all memory arrays will have "black-box" control on outputs that feed scan logic—all inputs will be fed to scannable monitor flip-flops)
  - No transistor-level design constructs or constructs that can't be modeled as gate-level elements
  - No synchronous latches
  - No transparent latches
- *Shift Stability*
  - No non-test logic gates in the scan data paths or scan control logic
  - No asynchronous functions or functions that have higher sequential operation priority than scan shifting

- No multiple clock domains within a single scan domain
- No internally generated clocks during scan (the test clock must come from an external pin to be accessed directly by a tester)
- No derived clocks (negedge or frequency division)
- No gated clocks (no clock signals gated by data, functional state, or other clocks)
- No clocks used as data (for example, a pulse generator)
- No data used as clocks (for example, a ripple counter)
- No data inversion between scan cells
- No shift "minpaths" (the clock-to-out and wire route propagation time delay between any two flip-flops should be greater than the maximum clock skew between any two flip-flops)

- **Scan Efficiency**
  - No long scan chains (scan chains should be less than 150 bits)
  - No unbalanced scan chains (all scan chains should be the same length)
  - No non-coincidental scan operation (all scan chains should operate simultaneously—no load and park operations)

- **Scan Frequency**
  - No slow scan interface connections (scan inputs and scan outputs should have better input setup and clock-to-out timing than the functional signals)
  - No extra logic added by scan insertion in a timed gate-level netlist (scan insertion should just add scan signal ports—SDI, SDO, SE—and internal scan routes—SDIs, SDOs, SE—and swap non-scan sequential elements for scannable sequential elements; No additional gates, Muxes, or sequential elements, these should be created in the HDL/RTL if needed)
  - No slow SE (SE should be routed as a clock-tree if necessary)
  - No registered scan control signals (sequential elements in the scan control logic result in moving scan signal analysis and consideration outside the ATPG tool sample cycle and into the shift operation analysis and consideration, and may limit the needed multiple-cycle assertion for path delay, transition delay, and other multi-cycle algorithms)

As can be seen, this set of very restrictive scan rules exists for several reasons:
- The ATPG tools that are used to create scan vectors can't deal with certain logic constructs or timing (ATPG tools are generally zero delay or unit delay in their consideration of vectors—so timing information is not considered by the tools).

- The ATPG tools require deterministic circuit behavior within each defined sample clock cycle, so pulse generators, combinational feedback, cross-coupled gates, X-generators, and Z-generators may result in vectors that are not valid.
- The act of scan shifting may be destructive if not managed properly—shifting is generally not seen or understood by the ATPG tool—any structure that may create driven contention (a node being driven by two drivers to opposite logic values) when random values are shifted through the scan chains can damage the final silicon during the testing process.
- Making sure that the scan chains can operate—gated clocks, derived clocks, derived data used as clock signal, asynchronous sets and resets, scan mode select logic, and non-test gates in the scan path connections can interfere with the shift operation or corrupt delivered scan data.
- Making scan easy to create, manage, or insert/install by not requiring an overly complicated design flow—mixing scan and non-scan design elements, dealing with multiple clock domains and negative-edge clock domains, mixing latch-based design with flip-flop-based design, and asynchronous input pin to output pin paths all complicate the scan methodology.
- Limiting the frequency of scan operation can limit test coverage and test cost goals—no "designed in" restrictions to the potential scan frequency should be allowed.

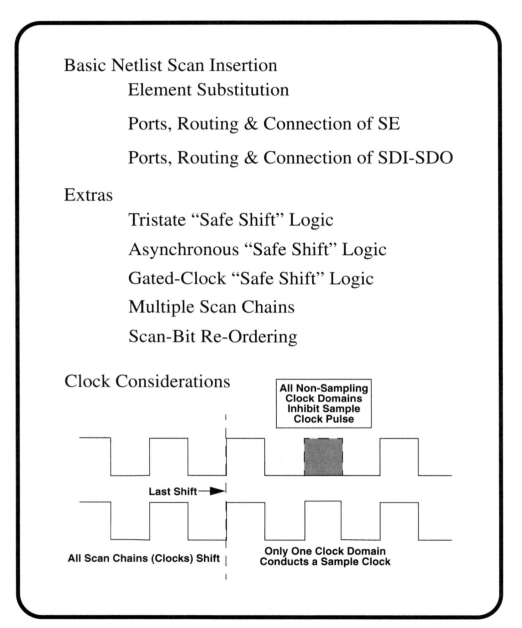

Figure 3-17 DC Scan Insertion

3.17 Stuck-At (DC) Scan Insertion

3.17.1 DC Scan Insertion

If a scan architecture is to be created that supports only the DC fault models, then scan insertion (see Figure 3-17) can be accomplished in just one or two steps: gate-level netlist element substitution and scan signal stitching. An optional third step may be supported to minimize scan routing—scan chain re-ordering in the physical place-and-route tool. Since no AC fault model requirement exists, and scan operation can be applied at any "slow" frequency, then the scan architecture does not need to meet any aggressive timing criteria. The scan operation can run at either the limit of the scan chain itself, the limit of the longest path in the chip (the frequency limiting path), or the limit of the tester—whichever is the slowest operation.

3.17.1.1   Element Substitution

The first operation in DC scan insertion is to substitute the regular flip-flops with the scannable flip-flops. This is normally done in the gate-level netlist. Ideally, these scan flip-flops have the same functionality as the non-scan flip-flops, and the scan flip-flops should be made with safe-shifting as a priority—no asynchronous functions or synchronous functions with higher priority than scan shifting.

3.17.1.2   Scan Enable Routing

The scan enable signal that toggles the multiplexor select line on all scan flip-flops needs to be assigned to a package pin port, and it needs to be routed and connected to every scan flip-flop in the design. Since no frequency requirement exists, the signal does not need to be fanout-treed for timing reasons (it may be treed if doing so is an advantage in managing the metal routing of the signal, or if signal skew is a concern).

3.17.1.3   Scan Data Signal Routing

The scan chain data connections (the serial stitching of SDI to SDO between the flip-flops) need to be routed to make the scan chains. The beginning and ending of the scan chains also need to be assigned and connected to package pin ports as well. These actions can be accomplished on the gate-level netlist. However, if an arbitrary routing uses an excess of metal, or if it prevents the router from completing, then a more optimum set of connections can be routed based on the physical placement locations of the design's flip-flops.

3.17.2   Extras

For DC scan insertion, where timing criteria is not a concern, the "safe shift" features can be gate-level netlist inserted. Other extra features that can be applied during scan insertion are the ability to define and connect multiple different scan chains in the gate-level netlist; the ability to balance the scan chains (same number of bits per scan chain—with only one odd short chain if there are leftovers), the ability to re-order the scan bits within a scan chain or between scan chains, and the ability to use the Q or QB functional signals as the SDO connection. The optimal re-ordering techniques can be applied by the physical layout tools, but the re-ordered scan chain information may need to be back-annotated to the ATPG netlist. This feature should also be supported by the scan insertion tool.

3.17.3 DC Scan Insertion and Multiple Clock Domains

A common analysis that must be applied after scan insertion is to verify that "clock versus data" races do not occur for the scan shift connections. If the scan architecture was assembled with a nearest neighbor type of algorithm, then the SDO-to-SDI shift connections are from a flip-flop to the next physically nearest flip-flop. These connections can be very short, and the physically nearest flip-flops aren't always on the same branch of the clock tree. This possibility makes the scan shift architecture very sensitive to clock-skew.

For slow, or DC, scan operation, the clock frequency, or data rate, of the applied scan test vectors is not the rated frequency of the part. All clock domains may be operated at the same slow frequency, but some care must be taken not to exercise any "hold time" violations or "clock-data" races when operating all clock domains during the sample mode (these timing problems are frequency-independent). Scan can exercise paths that are "false" in the sense that they are not true functional states; however, scan can install any state, even non-functional states; so all paths that can be exercised with scan are true paths during scan mode. Sometimes these false paths are not considered in timing analysis, and when the clock architecture is assembled, excessive clock delay or clock skew may occur between any two clock domains.

The safest course, in the presence of excessive clock skew between clock domains, is to load all scan chains at the same time but constrain the ATPG tool to allow only one time domain to conduct the scan sample. The only way to avoid this type of inefficiency is to ensure that the chip has been adequately time managed in clock skew and delay between clock domains (and all paths, including false paths, are analyzed for timing). A good "rule of thumb" for scan-based designs is to manage the clock skew to be less than the clock-to-Q time for the fastest scan flip-flop.

Most vector generators contain no timing assessment capability. So each scan or time domain is treated identically during vector creation. The installation of the proper timing relationship generally occurs in the vector writing or vector translation step. If clock skew or clock-data race conditions are suspected, then the scan vectors should be simulated with timing before being submitted to the test program.

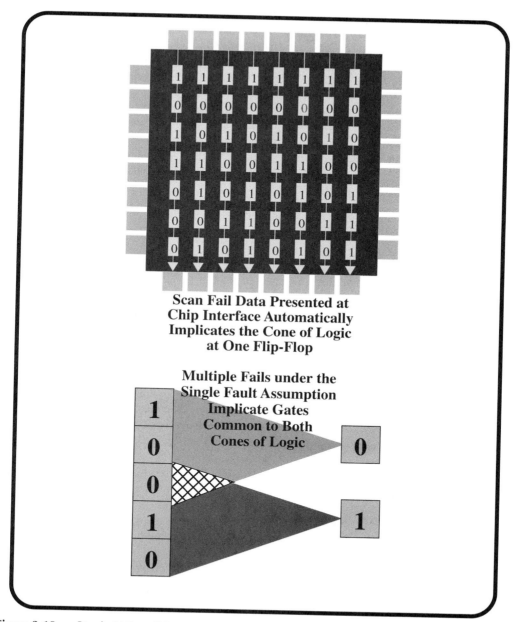

Figure 3-18   Stuck-At Scan Diagnostics

3.18 Stuck-At Scan Diagnostics

3.18.1 Implementing Stuck-At Scan Diagnostics

A scan architecture allows some very powerful diagnostic and debug techniques to be applied. Debug, diagnosis, and isolation are critical when the initial silicon fails wholesale (a massive number of failures) or when a fabrication facility starts having massive failures on parts that were previously passing. This type of "process monitoring" is known as "yield turnaround" and concerns the defect content imparted to the design during the manufacturing process. Another key concern is when a customer returns a product and the manufacturer must determine whether the failure is due to a reliability issue (defect-driven) or a latent bug (design error).

The two most common scan techniques used to help with these issues are diagnostic fault simulation and functional scan-out.

3.18.2 Diagnostic Fault Simulation

For slow, or DC, scan methodology, a fail indication on the scan data exiting the tester has the remarkable property of mapping directly to a flip-flop, because each clock cycle of a scan test program is associated with a shift-bit. A vector being scanned out of a scan-based chip relates each scan flip-flop to a bit of functional sample data—in other words, all the "cared" data coming out of the scan chains is data sampled into specific scan chain bits during the sample cycle. As a result, a failing scan bit maps directly to all the exercised faults within the cone of logic that drives data into the individual scan flip-flops (see Figure 3-18).

A powerful scan diagnostic technique can be applied to isolate the specific gate elements that may contain the defect. The technique involves taking multiple scan failures within one scan vector and across several scan vectors, and then fault simulating just the failing vectors (and more specifically fault simulating just the failing bits). The fault simulation is done without fault dropping so that every fault implicated by every failure is evident. Then an analysis can be applied to see which exercised faults appear repeatedly in the collection of failing tests. The collection of faults, or single fault, that appears in most of the listings is the most likely candidate as the cause of the problem.

In an ideal sense, the diagnostic fault simulation should be accomplished under the "single fault assumption" that the vectors were generated against, but sometimes this restriction will never produce an "exact match" for one single fault. This is because "real" defects usually do not follow the single fault assumption and evidence themselves as just one fault—real defects may take out several transistors, gates, wire connections, and circuit elements. To guard against this ambiguity, a useful diagnostic fault simulation should be based on a probability analysis.

For example, if a set of scan tests fails several individual bits within 1 vector and several individual bits in 3 other vectors, then a total of, let's say, 16 fails may cover 8 different scan bits. If all 16 fails, covering the 8 bits, are fed to the diagnostic fault simulator with the directive to find one fault that caused all the fails, then the analysis may never converge. To get the tool to converge would then require the user to start removing vector fails from the list to see whether some combination of fails finds an exact mapping to one fault. Even for the case of 16 fails covering only 8 bits, the number of permutations involved is a considerable amount of work on the user's part.

3 • Scan Architectures and Techniques

To make the diagnosis more "realistic" and usable, the directive of the diagnostic fault simulator would be to report the multiple faults that appear the most in the collection of faults and to rank them. For the case of 16 fails covering 8 bits in the scan chains, then the faults would be associated with the 8 cones of logic, and the only faults that would be kept in the analysis would be the ones that show up multiple times. So one fault may be associated with 12 fails, one fault with 5 fails, and so forth.

3.18.3 Functional Scan-Out

The other powerful diagnostic technique that can be applied requires supporting a combinational-only scan mode control, because the scan-out technique is based on functional vectors instead of scan or ATPG vectors (a customer is more likely to send a bad part back with a piece of failing functional code, not a scan vector). The technique is to apply the functional vector, and then on some identified clock cycle, to switch into scan mode and scan out the state of the part (and this can be done repeatedly on every functional clock cycle if necessary).

The scan-out technique allows the state of the chip to be viewed exactly as a simulator views the gate-level netlist. In fact, the power of the scan-out technique is to run the functional vector in conjunction with a functional simulation and to identify when the two representations differ in any state element. Once a difference is identified, ATPG-based vectors can be generated and used with diagnostic fault simulation to identify whether the failure is a defect or a bug (no difference between the structure of the netlist and the structure of the silicon indicates a functional bug).

BASIC PURPOSE

- FREQUENCY ASSESSMENT
- PIN SPECIFICATIONS
- DELAY FAULT CONTENT

COST DRIVERS

- NO FUNCTIONAL VECTORS
- FEWER OVERALL VECTORS
- DETERMINISTIC GRADE

Figure 3-19 At-Speed Scan Goals

3 • Scan Architectures and Techniques

3.19 At-Speed Scan (AC) Test Goals

3.19.1 AC Test Goals

There are two basic types of testing: *DC testing*, which is done to verify structure independent of frequency or timing, and *AC testing*, which assesses frequency and timing criteria (see Figure 3-19). The AC testing is needed to verify frequency compliance, pin specification compliance, and the manufacturing-induced delay defect content.

3.19.1.1 Frequency Verification

Frequency verification is accomplished by ensuring that all critical timing paths in a design actually make their specified timing in silicon after manufacturing. More specifically, the one worst real register-to-register transfer for each flip-flop must occur within the rated cycle time.

3.19.1.2 Pin Timing Specifications

Related to frequency verification is pin timing specification verification. Pin timing is the timing involved with the chip pin interface for such elements as: input-setup, input-hold, output-valid, tristate-valid, and output-hold. These specifications are verified by ensuring that the one worst-case longest path, or the one best-case shortest path, per pin, falls within the timing zone listed in the specifying document (a data sheet or data book).

3.19.1.3 Delay Defect Content

Any delay defect content that may have been added to the design during the manufacturing process must be assessed. The methodology to identify whether delay defect content exists and whether it can cause chip failures is to exercise critical paths such as those for frequency and pins specification timing, but with more "critical path depth" per endpoint (this means that more than one critical path per output pin or per register has a vector to verify the timing—this is also known as "peeling the timing onion").

3.19.2 Cost Drivers

The reason to conduct AC verification with scan is to better meet the "cost-of-test" goals. The cost drivers and advantages associated with scan-based AC testing are: shorter test times, eliminating functional vectors, fewer overall vectors (the scan vectors are more efficient in size and coverage), and a deterministic grade.

3.19.2.1 Shorter Test Time

The definition of AC scan testing does not mean to apply the complete scan vector at speed. All that is really required is that the sample interval is conducted at the rated clock frequency. However, if the shift and the sample are conducted "at-speed," then a test time cost reduction occurs naturally. For example, if 10,000 scan vectors that are "99 shifts plus 1 sample" clocks deep (100 clocks each) are run at 1 MHz, then the 1 million clocks associated with this scan test will be applied by a tester in a 1-second time period. If these same vectors are applied at 10 MHz, then it is a direct division of test time by 10—this means an applied time period of .1 second (100 milliseconds). If these same vectors are applied at 100 MHz, then this would be a direct test time division of 100—a test application time period of .01 seconds (10 milliseconds).

This reduction allows more tests to be conducted in the same amount of time, or allows a higher throughput rate on the production test floor. Note, though, that the reduction of test time will hit a "point of diminishing returns" when the test time is the same as the handler sequencing time (the speed at which a part can be inserted into and removed from the tester socket). For example, a common test time involved with modern IC handlers on ATE is about three seconds—if this is the handler supported, then there is no real advantage in reducing test time below three seconds.

3.19.2.2 No Functional Vectors

Scan vectors have been shown to be more efficient than functional vectors (in most cases, though exceptions exist). If they are not efficient when made, then algorithms can be applied to compress the scan vectors rather easily if the chip or package power rating allows. Efficiency means more fault coverage per clock cycle (which is a measure of lower cost-of-test). Since scan vectors will be used for DC structural testing, then the application of more scan vectors for AC purposes is just an extension of the existing test methodology (the AC and DC vectors have the same look and feel). Scan vectors are generally applied with a single and a simple edge set per pin, whereas functional vectors usually require complicated timing edge sets (which can lengthen the test program needlessly).

3.19.2.3 Fewer Vectors

Since scan vectors are more efficient, then fewer scan vectors are required for a higher level of coverage. The goal is to get the highest possible coverage with the minimum number of applied clock cycles (and to apply them at the highest possible speed).

3.19.2.4 Deterministic Grade

Currently, no "fast" timing assessment simulator is commercially available for conducting a simulation of vectors specifically for frequency and pin specification verification (one that can grade a set of vectors for speed path coverage). The timing simulators that are available do not grade "critical paths," or are not applied at the higher level models (for example, a low-level timing simulation would be a spice simulation). AC-based path delay scan vectors, however, are generated against critical paths. This means that path delay scan vectors are deterministic in coverage when they are created (as compared to functional vectors, which target behavior but may not target specific critical paths).

3 • Scan Architectures and Techniques 147

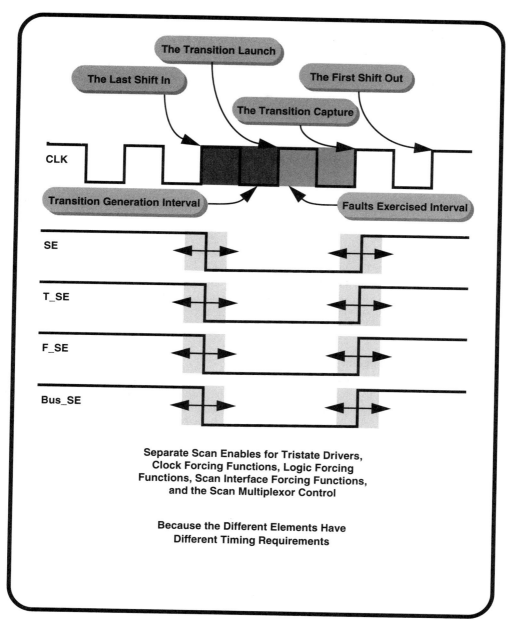

Figure 3-20 At-Speed Scan Testing

3.20 At-Speed Scan Testing

3.20.1 Uses of At-Speed Scan Testing

At-speed testing is the application of vectors at frequency (applying stimulus and evaluating response with respect to the various applied clock signals). AC scan testing is the application of scan sample interval (or intervals) at the rated clock frequency (see Figure 3-20). Conducting the shift process at-speed is optional but is more cost effective and is used for test time reduction. The term "at-speed scan" actually refers to applying the shift and sample of scan vectors at the rated frequency, but it is largely misused as a synonym for "AC scan." Note, however, that some ATEs are restricted to applying the same clock between the shift and sample operations (they have no way to change clocks or the clock period), so both the shift and the sample operations must occur at the same clock rate.

3.20.2 At-Speed Scan Sequence

Another difference between the DC scan operation and the AC scan operation is that AC scan requires vector pairs for certain operations. Vector pairs can be applied with the last-shift method or with the two-cycle sample. The last-shift method requires shift-bit independence, which negatively affects the scan chain stitching process (every even flip-flop in the scan chain must have nothing to do with the cones of logic of the odd flip-flops—the scan chain route may zig-zag severely to meet this requirement). The two-cycle method places more reliance on the ATPG software to create vector pairs with two consecutive sample (functional operation) intervals.

3.20.3 At-Speed Scan versus DC Scan

The difference between DC scan and AC scan is that the scan sample interval must occur "at-speed." Therefore, all scan enable signals and scan "safe shift" control signals must transition and their effects must resolve in much less than a clock cycle (so the scan signal transition is not what is being measured as establishing the timing specification). This requirement also means that any borrowed scan interface signals, if they are to be tested, must also resolve to the functional state in less time than the rated pin specifications (again, scan control can't establish the timing rating).

In DC scan insertion, all the scan control signals and scan enable signals could be inserted by the gate-level netlist scan insertion tool, because the scan chain is intended to be operated at a slow speed relative to the clock frequency. However, for at-speed scan, the scan enables and the "safe shift" control signals must resolve as aggressively as system logic. Since each of the scan control signal environments has different timing requirements, the best approach is to support multiple scan control signals.

For example, the scan interface control signal deals with interface multiplexors and pin tristate drivers; the internal "no shift contention" signal deals with internal tristate drivers; the "block asynchronous functions," "block higher priority functions," and "force set-reset on non-scan logic functions" type of signal deals with logic; the "disable clock gating" signal deals with clock nets; and the scan shift-sample control signal deals with the scan multiplexor on the scannable flip-flops. Each of these environments operates under different timing conditions, and so each should be controlled with its own signal.

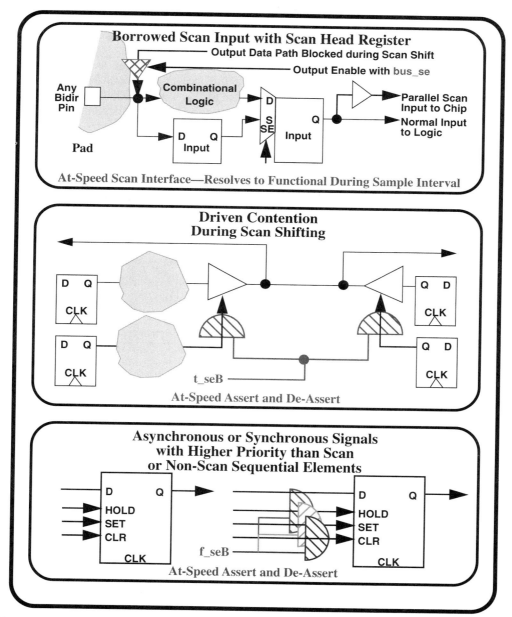

Figure 3-21 At-Speed Scan Architecture

3.21 The At-Speed Scan Architecture

The ideal at-speed scan architecture will allow both the shift operation and the scan sample operation to occur at the rated frequency (see Figure 3-21). This allows vectors to be generated to meet the AC goals, and it reduces test time by loading the scan chains at a higher frequency. There are only two roadblocks to applying the ideal at-speed scan architecture: making a scan architecture that operates at speed and the chip's power limitations (an at-frequency toggle rate may exceed the chip's power limitations or the package's thermal limitations).

3.21.1 At-Speed Scan Interface

The scan interface must take data from the tester and must deliver response data to the tester with the same timing that the functional interface would. This action allows the measurement of pin specification to be done with scan. Pin specification testing can be accomplished by providing a mechanism that mimics the functional interface, which can capture new applied scan data on inputs, presents internal scan data on outputs, and does not block or interfere with the functional interface. One method used to manage this measurement is to have dedicated Head and Tail registers in the scan interface, and to control the input versus output data direction with a dynamic scan interface enable signal (e.g., bus_SE).

3.21.2 At-Speed "Safe Shifting" Logic

The "Safe Shifting" logic installed in the chip design to control tristate busses, asynchronous logic, gated clocks, uninitialized sequential elements, and other similar conditions must also be controlled by dynamic at-speed control signals (for example, t_seB and f_seB in Figure 3-21). These signals can be installed in the model and synthesized with timing constraints to ensure that they do not limit the speed of the scan operation.

3.21.3 At-Speed Scan Sample Architecture

One of the fundamental requirements for at-speed scan sampling for vector pair analysis and generation is that the circuit must resolve to an almost pure functional state. This is because vector pairs are state-nextstate vectors—and blocked, tied, redundant and constrained nets, nodes, and gates can result in invalidating many possible state transitions.

For example, one of the common techniques to use in DC stuck-at scan architectures is to fix the outputs of the embedded memory arrays to a known value and to capture the input side with monitor registers (generally referred to as black-boxing the memory). However, this technique may prevent the cache memory from interacting with the tag memory, and therefore signals like cache-hit, tag-valid, dirty-data, etc., may not be generated during the scan test mode. These signals may be used as global signals that feed into many cones of logic and allow the flip-flop endpoints to update. If these signals are blocked, or fixed to one logic value, then entire cones of logic may not be able to conduct state-nextstate updates. This artificial test architecture may eliminate many valid critical paths from ATPG analysis. Therefore, these types of techniques and practices should be avoided. Refer to the scan memory interaction sections 4.14 to 4.18 in Chapter 4.

3 • Scan Architectures and Techniques

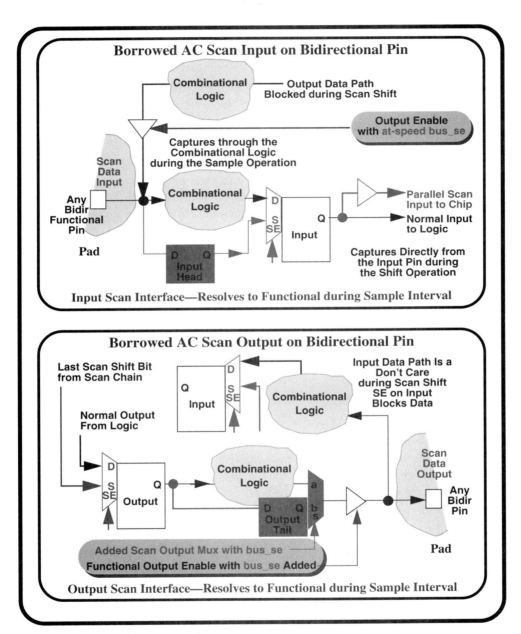

Figure 3-22 At-Speed Scan Interface

3.22 The At-Speed Scan Interface

The ideal at-speed scan architecture will allow both the shift operation and the scan sample operation to occur at the rated frequency, and should allow the scan interface to be used to verify the AC pin specifications. These actions can be done only if the scan interface allows the transition from shift to sample (functional interface) and back to shift to occur at the rated frequency (see Figure 3-22).

3.22.1 At-Speed Scan Shift Interface

The scan interface must take data from the tester and must deliver response data to the tester with the same timing (edge set) that the functional interface would. This action allows the measurement of pin specifications to be done with scan and can be accomplished by providing a mechanism to capture scan data on inputs and to present scan data on outputs that does not block or interfere with the functional interface. One method used to manage this is to have dedicated Head and Tail registers in the scan interface (see Figure 3-22). Scan Heads and Tails ensures that the beginning and ending registers of the scan chain can be placed close to the shared package pins used with the scan interface. The dedicated registers should be modeled and passed through the synthesis process with timing constraints to ensure at-speed operation. To support at-speed scan, the internal flip-flops must also operate at speed. So scan data routes and the scan enable (SE) must be capable of operating at the rated clock frequency. Some form of route management must be applied (for example, a nearest neighbor scan data routing order).

3.22.2 At-Speed Scan Sample Interface

If functional pins are borrowed as the scan interface (scan inputs and scan outputs), then these scan signals must allow the pins to resolve to their functional purpose during the functional sample interval. An at-speed scan architecture requires that for the transfer "from scan Heads and to scan Tails," to and from, the functional registers must occur at the rated chip frequency. This means that scan inputs and scan outputs borrowed from functional input pins, output pins, and bidirectional pins must have a dynamic interface control signal.

Allowing the scan interface to resolve to the functional state allows the pin specifications to be measured on the pins borrowed for the scan interface. The control signal (generally referred to as bus_SE) is a dynamic scan enable type of signal that can be modeled and passed through the synthesis process with timing constraints. It is a separate type of signal from the SE used to control scan flip-flops—this is why it should be a separate signal with its own input pin.

Note: The dedicated scan Head and scan Tail registers allow the first and last bits in the scan chains to be "functional Xs." This is important, since the functional register, if used as the first element in the scan chain, may have its capture data from the input pin (the pin specification) overwritten at the output of the scan chain if the last shift out is used to measure the output pin specification. Note that Head and Tails do not need to be scannable flip-flops.

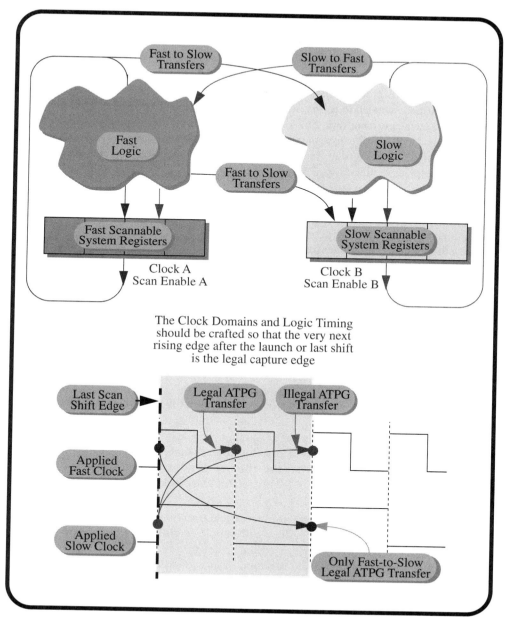

Figure 3-23 Multiple Scan and Timing Domains

3.23 Multiple Clock and Scan Domain Operation

3.23.1 Multiple Timing Domains

In the DC scan environment, where timing is not considered, multiple time domains are usually handled by scan inserting all logic identically. Optionally, the scan chains may be separated (scan inserted) by clock domain because of clock skew and clock-data race concerns, but the method of use is intended to be operation of all scan chains at the same test frequency.

However, whenever multiple clock domains exist and the scan operation needs to take advantage of the clock relationships to meet AC test goals, then the ability to scan shift each time domain and sample from one time domain to another, with respect to the actual clock, timing, or phase relationships, is required (see Figure 3-23).

Two or more time domains can be supported for scan testing, as long as the registers that make up the scan chains are separated by clock domain. This means that there can be no intermixing of scan elements connected to different clocks within one scan shift chain, and each scan timing domain must have its own separate SE signal. Separate "scan domains" (by time domain) allow each time domain to be scan shifted and sampled separately, and with the proper clocking (phase) relationship to each other.

The basic rule for successful ATPG of separate time domains is that the logic in the "cross domain transfer" zone should be timed to the "next rising edge" clock. The launch of scan data is based on an edge in one clock domain, and the capturing register operation is based on a different edge in a different clock domain—the capture should be the first clock edge at the correct phase relationship after the launch in the target or capture clock domain (not the next rising edge of the launch domain). The key is to apply a timing condition to the logic involved in the data transfer between the two domains and to then be consistent in applying this constraint to the design. For example,

- All fast-clock-to-slow-clock transfers (the logic between the two clock domains) must meet slow cycle time timing—this happens automatically since the slow capture clock happens at a later time than the next fast clock. Note that the fast clock may have to skip any clock pulses that happen before the slow capture clock to ensure that the change of application data does not compromise the conditions of the test (or the fast clock can be delivered from the tester as identical to the slow clock—see Figure 3-23).

- All slow-to-fast transfers must meet fast cycle time timing. This means that no "fast clock domain" scan flip-flops will sample "multi-fast-cycle paths." The timing of all logic between the slow domain and the fast domain must pass data at the cycle time of the fast clock period. If this condition is violated, then the fast scan elements that sample slow paths must be forced, or constrained, to capture only the "X" logic state during all scan operations that occur at the "fast" frequency, or the fast time domain must be operated at the slow clock frequency during test.

As another example, taking the case for two time domains with one clock at half the frequency of the other (see Figure 3-23), if both clocks have simultaneous last shift edges and the data is launched from a slow clock scan register to a fast clock scan register, which edge is the correct capture point? The fast clock will have two clock rising capture edges within the slow clock sample interval (conversely, the fast clock will also have two launch edges if the slow clock register is used for capture). The logic should be made so that the first rising edge after the last shift (in either clock domain) should be the valid capture edge. Some tricks may be played on the tester, such as operating all the clocks at the slow speed if all inter-clock domain data transfers are accomplished based on the slow clock period (the logic between the two domains is designed to operate at the slower timing target). Since the data propagation time is targeted to be at the slow period, operating both the fast and slow clock at the same slow frequency does not compromise the testing—the logic will be verified at the correct timing interval.

Vector generation for a multiple clock domain type of architecture may be done in separate groupings to ensure that the fault coverage is isolated by time domain. For example, the fast time domain is vector-generated while discounting the slow domain, the slow domain is vector-generated while discounting the fast domain, and the cross domain transfers are done with the fast domain being the capture registers for slow domain launches and the slow domain being the capture registers for fast domain launches. Sometimes vector generation separation is necessary, since the ATPG tool will require a setup, procedure, or constraint condition to withhold any clock pulses.

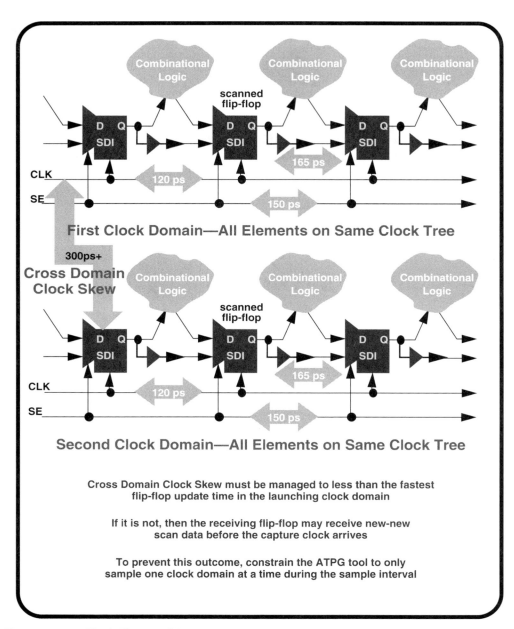

Figure 3-24   Clock Skew and Scan Insertion

3.24 Scan Insertion and Clock Skew

3.24.1 Multiple Clock Domains, Clock Skew, and Scan Insertion

One of the concerns involved with scan insertion is "scan shift data races." These occur because the scan architecture is basically a direct connection from one flip-flop to another (Q or SDO to SDI)—if the data transfer during scan shifting, from one flip-flop to another, occurs more quickly than the maximum clock skew associated with the clock architecture, then there is a probability that the data in the first flip-flop will change and that this changing data will arrive at the second flip-flop before the clock will. If this outcome occurs, then the second flip-flop will receive incorrect data, and the old data that was contained in the first flip-flop vanishes—this effect is sometimes known as *data smearing* (new data passed through the first flip-flop and ended up in the second flip-flop).

Data smearing can occur within a single time domain if the clock-to-Q of any flip-flop, and the propagation delay to the next flip-flop, approaches the maximum skew associated with the clock tree. This problem only gets worse when multiple time domains are involved—for scan testing, not only must the skew timing of the clock architecture involved with one clock domain be managed, but the skew timing involved with other clock domains must also be managed. Note that in deep-submicron designs, clock skew management is becoming a very important design issue—in some .25 micron designs, the flip-flops have clock-to-Q timing near 100ps, and clock domains with thousands of flip-flops have maximum skew tolerances greater than this (e.g., 150ps). If a nearest-neighbor scan optimization strategy is used by a place&route tool, then the result may be a worst-case "shift skew" design.

The management of the "shift race" or "shift skew" problem can be accomplished in several different ways: 1) by globally ensuring that the maximum clock skew associated with any given clock tree is less than the clock-to-Q of the fastest scan flip-flop in the scan architecture; 2) by globally ensuring that the data connection propagation timing from any one scan flip-flop to another is greater than the clock skew rating of that clock domain; 3) by evaluating the individual scan connections with reference to clock skew and fixing each identified problem area (adding delay buffers, changing the drive strength associated with the SDO, and swapping scan data connections with other scan elements to make longer SDI-SDO routes).

A warning must be given here for the DC scan test architecture—if a design with multiple clock domains (multiple clock trees) is to be tested without AC scan considerations (all testing is to be done at one test frequency), scan separation into clock domains may still be required. If each clock domain is skew managed but the interaction between clock domains is not skew managed, then placing flip-flops from any time domain (clock tree) onto a single scan chain can result in a nightmare of a scan shift operation (the clock skew for this architecture is the worst-case clock skew between the two clock trees). This type of design may occur if the scan insertion engine acts on a design description that "ties" all the clock domains together to one test clock—then the tool mistakenly believes that there is only one scan domain on one clock domain. Even for DC operation, it is best to separate scan chains by clock domain, and to provide individual control for each scan domain.

In multiple clock domain test architectures, especially for AC scan testing, each clock domain must be scan inserted as a separate scan domain, and each clock domain must be clock skew managed to eliminate "shift skew." In addition, the cross clock domain test transfers should be skew managed to prevent "sample skew."

As was stated previously, the logic between any two clock domains must be constrained to meet the data transfer timing of one clock domain or the other. There is more involved than just the logic making timing; the launch and capture clocks must also not have "clock skew" problems within and between the clock domains. For example (see Figure 3-24), one clock domain/scan domain may have a maximum skew of 120ps and the clock-to-Q of the flip-flops is 165ps—there is no "shift skew" problem with this scan domain. A second clock domain/scan domain may also be timing managed identically—there is no "shift skew" problem with this scan domain. However, a normal operation of loading both scan chains and then conducting a "simultaneous sample" may not be successful if the skew between the two time domains is excessive.

What may happen if the update time within one clock domain is very fast and this time is less than the maximum clock skew rating between the two time domains is that the target time domain may capture new-new scan data. This problem is very similar to the "shift race," except that it happens during the sample cycle—a flip-flop in the launching clock domain successfully samples during the sample cycle, and the new data arrives at the target capture register before the sample clock in that time domain arrives (or before the target capture register's hold time).

From a design point of view, many designers will say, "But this is the functional operation and it must make timing." This condition may create several problems for scan, for example:

- the scan may activate "functionally false" data transfer paths, or
- the scan update in the launching time domain may be based on "functionally false" and short paths, or
- the scan clocking sequence may be different or based on a different clock source from functional operation.

A scan architecture allows any state to be placed in the scan chain, and so scan may enable pathways not exercisable by functional operation.

The two commonly applied solutions to the "sample skew" problem are: 1) manage all the clock domains to meet timing goals as if they were one large clock domain during the scan test operation; 2) constrain the ATPG tool to update only one scan domain at a time during vector generation. Meeting solution #1 is getting harder and harder as process geometries shrink—and the work involved is sometimes considerable, which can negatively affect the schedule. Using solution #2 is becoming more common, but constraining the ATPG tool to update only one scan domain at a time means that the each scan domain can be clocked separately, which may result in several separate ATPG operations, more vectors, and increased test cost.

3.24.2 Multiple Time Domain Scan Insertion

One of the methods some design organizations use to minimize the test complexity is to limit the number of scan "clock" domains (after all, a bypass pin for each supported clock domain is an integration cost issue). It is an easy thought process, to want to combine clock domains that are similar (for example, to declare a scan test domain made up of 48MHz,

50MHz, and 45MHz clock domains because they are all very close in frequency of functional operation). This type of thinking means that the chip may support an actual ten different clock domains, but several of them are similar enough to consider testing at a single frequency. In this case, the number of resulting scan domains may be just three or four, whereas the number of clock domains is actually ten. Danger lurks in this approach (a hidden "work" hand grenade). Much more work may be required to "skew manage" the combined clock domains—even if they are close or identical in frequency (two separate 48MHz clock trees may be delay and skew managed internally, but are they also timing managed to each other).

One of the common mistakes made when crafting a multiple clock domain scan test architecture is to think of scan chains by the functional frequency. Scan chains may shift at any frequency, and they may sample as either a functional operation, or the timing involved with a data transfer from one clock/scan domain to another clock/scan domain may be crafted as a specialized tester operation. For example, all the scan chains in a device may be loaded at their natural frequency with great care taken to pad them with Xs so that the last shift occurs at the correct phase relationship for the two domains in question. A natural sample is then taken. This type of operation emulates the natural functional data and data transfers occurring within the chip, but this type of test requires extensive vector manipulation and some clever tester sequencing.

A much easier methodology for scan testing is to simultaneously shift data into all scan chains at some frequency (independent of each scan domain's natural frequency), where the last shift occurs simultaneously in all scan chains, and to then withhold the clock to all scan domains except for the one to be sampled—this allows a single scan domain to have a "stable sample" regardless of the skew management between clock/scan domains. If the test is for a 50MHz clock domain, then the shifting of all scan chains may be accomplished at 50MHz, and if the test is for a 48MHz clock domain, then the shifting may be accomplished at 48MHz. The requirement here is to allow the scan chains to shift at any frequency the test requires—not the frequency implied by the logic the scan chain services. A frequency-independent scan architecture is crafted during scan insertion and relies only on the power distribution and the timing involved with the scan data connections, the clock drivers, and the scan enable.

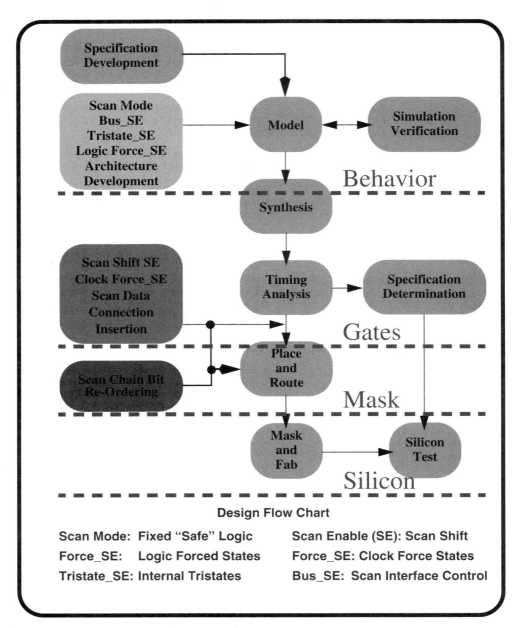

Figure 3-25   Scan Insertion for At-Speed Scan

3.25 Scan Insertion for At-Speed Scan

The scan insertion for at-speed is significantly different from the DC scan insertion method. Several of the scan insertion operations must now take into account the added dimension of timing. This complicates the scan insertion and distributes the scan insertion functions into three different design and build environments: the behavioral HDL or RTL model; the gate-level netlist (GLM); and the place-and-route (physical) database (see Figure 3-25).

3.25.1 Scan Cell Substitution

The scan cell substitution step is largely the same. There might be a need in the future to compare one scan cell to another for timing criteria, but this isn't currently done with commercial tools (for example, scan cell substitution and selection accomplished by timing optimization). Most scan insertion tools simply require that the substitution cell be stipulated by the user, or that it be stipulated in the library description of the non-scan cell—a scan equivalent is named within the description of the non-scan cell.

3.25.2 Scan Control Signal Insertion

All scan control signals that are used to establish "safe shifting" and to force or constrain logic such as memory array interfaces, the scan interface, tristate drivers, asynchronous elements, etc., must be dynamic to allow AC test goals to be verified for compliance. The connection of the scan enable (SE) and the dynamic forcing signals that operate similarly to the scan enable (for example, as described, the f_SE, t_SE, and bus_SE) must now be done with respect to critical circuit timing. In most cases the SE is treated like a clock in the physical place-and-route tool and is fanout-tree managed (skew and delay timing managed with some form of buffered routing). The other signals may be fanout tree managed, or they may be synthesized directly with timing constraints (and still fanout tree managed by the physical place-and-route tool, if necessary).

3.25.3 Scan Interface Insertion

The scan interface must also be given due consideration. The Heads and Tails are generally modeled so that they may be synthesized and timing analyzed to meet the interface specifications, but they may be netlist scan inserted as well (if the timing impact is analyzed). These cells should be physically optimized to be near the pins they service.

3.25.4 Other Considerations

The scan data routes must now also be at-speed operation rated. Since the scan connections do not exist in the model and since gate-level netlist scan signal insertion will not always meet timing criteria, the scan connections must also be managed by re-ordering in the physical database to meet some minimum timing or wire length.

If any of the scan control or scan data connections do not make (minimum or maximum) timing after they have been routed, then some extra work may need to be accomplished to upsize drivers and to buffer signals.

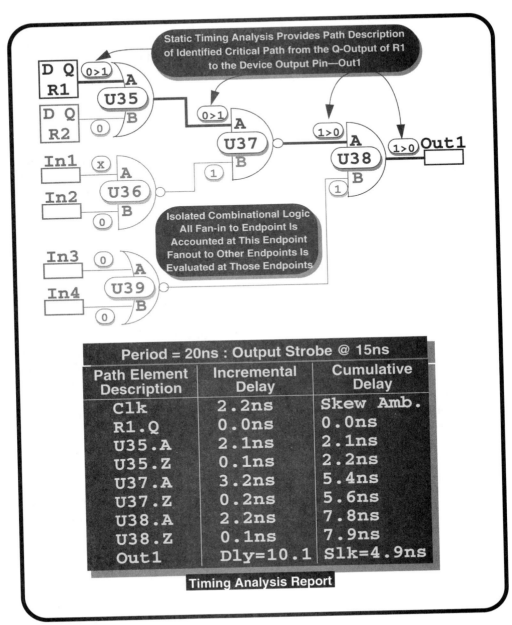

Figure 3-26 Critical Paths for At-Speed Testing

3 • Scan Architectures and Techniques

3.26 Critical Paths for At-Speed Scan

3.26.1 Critical Paths

The faults that are used for AC path delay testing should be complete critical paths. These paths come from some form of pre-silicon timing analysis, for example, static timing analysis that operates on the gate-level design description. The static timing analysis tool processes the design by evaluating the timing library against the collection of gates, the sequential elements, a wire-load model, and the clock distribution network. It produces a file with path descriptions, where a path begins on an input pin or a register, and ends on an output pin or a register—this action results in four classes of paths: input-to-register (I2R), input-to-output (I2O), register-to-register (R2R), and register-to-output (R2O). These path descriptions (see Figure 3-26) have the incremental delay that is associated with the gate elements in the described paths and the total delay of the "stack up" of incremental delay, and may also have the "slack" timing included (slack is viewing the path timing from capture end where delay is viewing the path timing from the launch end—slack is the time metric that the path propagation delay did or did not make it to the capture point by the setup time requirement—slack can be positive, "I made it with 200ps to spare," or slack can be negative, "I was late by 40ps").

These paths may be translated to a data format that can be submitted as complete entities to an ATPG tool as a path fault (the entire path is considered as the fault), or the elements of the path can also be input to the ATPG tool individually as linked gate delay faults.

3.26.2 Critical Path Selection

The faults that are used for AC path delay testing should be the timing analysis identified critical paths. However, the number of paths in a design is much larger than the finite number of gate elements and connections. Since a vast number of paths exist, as compared to individual gates and nets, a selection process must be established to create a reasonable fault set with a related vector set that would fit in the tester's memory (the economic considerations and restrictions of test time, no tester reloading, and limited tester memory). For this reason, only a budget of paths must be chosen or selected for ATPG.

The ideal path selection methodology is to know the test vector budget, to know the sizing of the stuck-at and transition delay vector sets, and to then pick the most critical paths up to the number that would fill the remainder of the vector budget. However, in reality, the job is not that easy. Several facts thwart this scenario: first, all the vectors are usually generated during the same time frame, so the other vector products' sizing are not known when path delay vector generation begins; second, the most critical paths in the chip are not always the best paths to select, since many of them may be in the same cone of logic; third, there is false path content in any static timing analysis file, so that many of the targeted paths will not result in vectors, or they will result in vectors for false paths; fourth, the timing analysis path descriptions are not usually organized in such a way that path selection is easy; and fifth, "What is the definition of a critical path anyway?"

To begin with, most static timing analysis tools rank all paths by slack (the amount of time that the path meets or exceeds its timing target) and/or propagation delay (the delay involved with propagating a logic value through the path in question). Most timing analysis tools present the results in a file of paths and in a histogram representation. In most cases, the tool is not asked to generate and then write every possible path into a data file (this task would quickly overrun many disks), but the tool is given a directive—only the top million, or only the worst path per every endpoint (flip-flop or output pin).

Two kinds of histograms can be used for the analysis: the true histogram, where paths are organized by timing from the absolute worst path (least or negative slack, or longest delay), and then by rank order for all other paths until the least worst path; or the "by endpoint" histogram, where timing endpoints are listed in rank order from worst to least worst, based on the one worst path that travels to that endpoint. Both descriptions have value, but the "by endpoint" is easier to use for filtering, since each endpoint is also a scan observe point—this allows vector creation that contains inherent diagnostic information. A concept of "critical paths per endpoint" can then be defined.

The AC test goals of a final silicon product are: 1) to verify the product's frequency; 2) to verify the product's pin specifications; 3) to assess the delay defect content added due to the manufacturing process. With these in mind, then the definition of a critical path would be:

Those paths that establish the product's frequency, or pin specifications, within some defined timing guardband, and those paths that would establish the product's frequency in the presence of delay faults that would not be caught by some other test, such as stuck-at or transition delay-based testing.

The best approach is to use critical paths for meeting AC goals, because they are very sensitive to delay faults and will mostly likely convert a delay fault (or an erroneous design timing library assumption) into a failure.

3.26.3 Path Filtering

Paths that are to be selected for ATPG to meet AC test goals are not always just the set of paths that have the worst propagation delay, or the least slack. To determine operating frequency of a part, then the paths involved with register-to-register transfers that are the single-cycle data transfer limiting paths should be selected. Note that only one path per register need be selected (in a defect-free environment, only the one worst path is needed to verify that the transfer is successful). To determine the pin specifications, only the one worst-case longest-input setup path and one best-case shortest-hold path per input pin; and one worst-case longest-output data, one worst-case longest-output enable, one best-case shortest-output data and one best-case shortest-output enable path per output pin is needed. Consideration of the path delay content increases the requirement to multiple paths per register and per pin.

Also note that some of the paths involved with frequency and pin specification determination may be so short in timing that they do not need to be included in the mix of paths that need vectors. For example, if the longest path from a particular register to another register was a direct wire connection and had a propagation time of 2 nanoseconds in a clock domain with a 20-nanosecond period, this is clearly not a frequency-limiting path and an 18-nanosecond delay fault would be required to make this path fail. However, for "specification" purposes, this vector may still be required.

3 • Scan Architectures and Techniques 165

Once the goals of critical path selection have been defined, then the path description database must be searched for paths that meet the criteria. For example, to build the input-setup pin specification path listing to feed to the ATPG tool, first all paths from input pins to internal registers that have a propagation delay above some timing target must be collected. Then this grouping can be further pared with items such as the clock domain target for the internal register, one path per pin for the pin specification assessment, an additional "N" paths per pin for delay fault content, whether the path is to the D input pin or to the CLR pin, and so on.

To assist in this effort, the path description must be understood. There are several components of a path description, and these components may be used as the filtering criteria for paths during the selection process. The path selection process is sometimes referred to as "path filtering," since path description file elements can be used to select, or filter, paths for use. Some path elements that can be used for filtering are:

Head   The beginning point of a path—usually an input pin or the Q of a register.

Tail   The ending point of a path—usually an output pin or the D, SDI, CLK, CLR, EN, or other input signal of a register.

Stream   Any element description within the path, except for the Head and Tail.

Delay   The total propagation delay of a path description.

Slack   The positive or negative number that represents the amount of time that the path meets or does not meet its time budget.

Clock Source   The launch clock involved with the Head.

Clock Sink   The capture clock involved with the Tail.

Path Type   A path description typing of register-to-register, input-to-register, output-to-register, or input-to-output that can be associated with the path.

3.26.4 False Path Content

The problem with static timing analysis is that it operates in the absence of boolean information (whereas ATPG operates in the absence of timing information). This results in tracings through the gate-level design descriptions that could never be exercised. There is a further subdivision of these paths into "They would never be exercised in functional mode, but maybe in scan mode" (since a scan architecture has so much more control and can install states in the design that would never occur with functional mode).

When paths that can never be exercised are submitted to the ATPG engine, the ATPG engine will fail to create a vector for them. If the reason for unsuccessful ATPG is investigated and found to be non-realistic path tracings, then these paths can be labeled "False" and ignored. In this manner, the ATPG engine can be used to filter out some classifications of false paths, such as: not boolean true; blocked-tied-redundant; and "no valid next state" for a state-nextstate type of path.

Note that any path that the scan architecture can exercise will be a "true" path for testing, even if it is a false functional path that can never be exercised by functional mode.

False path content can make an ATPG methodology for path delay seem to be an inefficient process. Thousands of paths can be submitted for ATPG, and only hundreds of vectors may be delivered. To the novice, this fact would be called low-fault coverage (hundreds divided by thousands results in just a few percent—for example, 100/14000 = 0.714%). The real answer is that the ATPG process has generated vectors for a "real" subset of paths, and the timing map is shaped to meet the targets and budgets involved with the "real" timing concerns.

3.26.5 Real Critical Paths

If all paths in the chip are considered valid for AC fault coverage, then the ATPG tool will seem to be very ineffective, since many of the false paths will not result in vectors. However, if the deterministic fault content is limited in some way, then the tool may be able to work to a finite fault coverage number. This can be done by selecting a lower bound or cutoff point in timing delay or slack. For example, a path with a 2ns propagation time between two system registers, with a clock cycle of 30ns, will have 28ns of slack—it would take a +28ns delay fault to convert this fault model into a failure. A defect that would cause a 28ns delay fault is a pretty significant defect that would most likely also be detected in the transition delay or stuck-at vector realm.

So a cutoff should be established that represents the sum of the largest delay fault and the largest library ambiguity that will not be detected by a different fault model. The current selection of this bound is still an inexact science (actually, it has been called a black art by some—the inexactitude exists because there is inadequate failure analysis information to describe exactly where a path delay fault converts to a transition delay fault and where a delay fault converts into a stuck-at fault.

In the absence of comprehensive process information, a good driver for the cutoff timing point would be the number of paths available versus the vector budget. Once the lower timing bound is established, the critical path histogram map should be used to drive ATPG to get one critical path per endpoint (where an endpoint is a system flip-flop or output pin) and a depth of paths behind each endpoint equivalent to the cutoff timing value. For example, if 130 endpoints have their worst timing paths above the timing cutoff bound, then these are the only "critical path" endpoints allowed. The next goal is to identify a number of paths within each of these endpoints' cones of logic that are also within the timing bound.

This type of "bounding" creates a box in the complete path histogram map. Once this is done, the false path content can be filtered out, and what is left is a box of "real" paths. The box now contains a finite number of elements. This box of "real" paths should be used as the denominator for "path" fault coverage.

3.26.6 Critical Path Scan-Based Diagnostics

Critical paths used as faults also carry inherent diagnostic information. A path description is a collection of gates, gate connections, and nets. In a similar manner to the DC scan diagnostic fault simulation, failing vectors can be related to their path descriptions (fault), and these path descriptions can be sorted and compared across several failing vectors to identify any common gates, gate connections, or nets. The difference between this analysis and diagnostic fault simulation is that the vectors do not have to be resimulated. The path delay vector has a target endpoint bit, and this bit is directly related to the path that was used to generate it, so no simulation is required (however, the location of the vector in the test program, the vector number, must be documented and tracked).

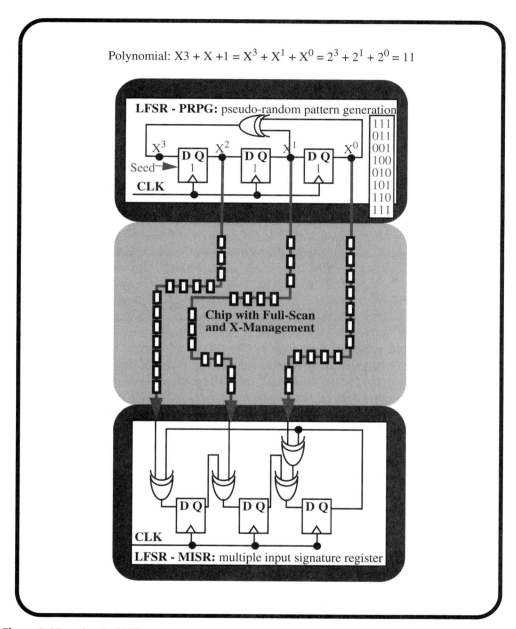

Figure 3-27 Logic BIST

3.27 Scan-Based Logic BIST

3.27.1 Pseudo-Random Pattern Generation

Sometimes the source of the scan vectors is not a tester, but an on-chip (or off-chip) counter of some sort. A common odd-ordered counting device used for random or directed vector generation is known as an *LFSR* (linear feedback shift register). An LFSR that is configured to generate a random vector stream is known as a pseudo-random pattern generator (*PRPG*). The PRPG-based vectors may be applied to an existing scan architecture directly from the taps of the LFSR, or through some decorrelation logic (logic that places some spacing between the data fed into different scan chains so that all chains do not receive the exact same data—it has been referred to in some instances as a "randomness randomizer").

There is another way to provide self-contained auto-vector generation, and that is to make the BIST logic part of the circuit. Instead of using a separate LFSR circuit, sometimes the scannable sequential elements themselves can be configured as LFSRs and can generate random pattern sequences "at speed" right within the scan chain domain (placing exclusive-OR feedback elements at certain places in the scan architecture to convert the scan shift architecture into a linear feedback shift register).

3.27.2 Signature Analysis

In a similar manner, sometimes the output verification device is not the tester, but an on-chip (or off-chip) device designed to capture the output and convert it to a "pass or fail" indication. One common device used for this purpose is the data compressor LFSR.

LFSRs that are configured to receive data and to compress the data stream into a single binary word are known as signature analyzers (response compressors). Signature analyzers can receive a parallel stream of data or a serial stream of data. Multiple parallel streams are processed by a Multiple Input Signature Analyzer Register (MISR), and serial streams are processed by a Serial Input Signature Analyzer Register (SISR).

In a similar manner to the embedded PRPG engine, instead of using separate LFSR circuits to conduct signature analysis, sometimes the scan cells within a circuit can be configured as signature LFSRs and can compress circuit responses "at speed" right within the scan chain domain.

3.27.3 Logic Built-In Self-Test

The use of LFSRs to conduct logic testing is generally referred to as Built-in Self-Test (*BIST*). When applied to general combinational and sequential logic, the configuration is termed Logic BIST, or *LBIST*. BIST is usually used only as a confidence test or as a method to reduce the deterministically generated vector sizing and can be done without reliance on a tester (for example, the test can be run when the chip is in the system) as long as a stable clock source is available (see Figure 3-27).

The test is generally used only as a confidence test for several reasons. One reason is that the PRPG LFSR is limited in the sequence of values that is created (all 2^n-1 values may be created, but not every permutation of sequences). If there is random pattern resistant logic in the design, then a particular LFSR with a particular polynomial may not achieve very high fault coverage levels (for example, an 8-input AND-gate requires a test where 7 input signals have a logic 1 applied and one signal has a logic 0, and in such a manner that the logic 0 input is detected uniquely—this may not occur easily or naturally—so this type of fault is termed *random pattern resistant*). To solve this problem a deterministic embedded vector generator may be used, but it currently requires a great amount of analysis to create, and generally carries a significant area overhead to implement.

A PRPG LFSR that is based on a primitive polynomial (certain prime numbers) will generate 2^n-1 states (all possible states but the all 0 state). If the circuit is fully combinational, then the full decoding of states may detect all testable stuck-at faults. A trade-off here is when the width of the combinational inputs is not too wide—a 32-bit wide LFSR is reasonable, but a 400-bit wide one is not; the larger the LFSR, the longer the test time—2^{400}-1 is a large test time).

BIST can also be used to exercise and detect at-speed fault models. Most timing faults require vector pair transitions. The PRPG LFSR is basically a minimal logic odd-order counter, and the count order depends on the polynomial, and the sequence order depends on the seed (the initial value in the LFSR). However, since the vector pairs are a function of the LFSR, and since there is a class of faults known as random pattern resistant faults (there are random pattern resistant faults for combinational depth of 1 as well as combinational depth of 2 or more), then the AC coverage can be very limited—even though the BIST logic and the circuit logic may be operating at speed. This is true for BISTs that drive scan chains, or for BISTs where the LFSRs for PRPG and signature analysis are the scan elements.

Note that one other type of LBIST exists—one where the LFSRs used for PRPG and signature analysis are applied to a chip without scan, or that scan is not used when the BIST is applied. This type of BIST is much like applying random vectors to the natural inputs of a chip, except that the LFSRs are not restricted to just being on the outside of the chip—they can be applied at any hierarchical level of the design if the design can support the extra logic and area involved.

3.27.4 LFSR Science (A Quick Tutorial)

The LFSR makes use of a boolean logic feature—multiplication is done by shifting left, and division is done by shifting right. If a shift register makes use of feedback from various bits while shifting right, then the operation is the division by two plus the addition of a constant. If the constant can be calculated to be a prime number, then the effect is that the shift register is dividing the input number by a prime number.

This outcome has the effect of always having a remainder unless the state of the register and the next added value is exactly the prime number. The PRPG operation will create all possible logic values in a register but the all 0 state, and the signature analysis operation always leaves the remainder in the register as the residual value. A polynomial that has these features is known as a prime polynomial or a primitive polynomial. The polynomial is where the feedback taps are taken from the shift register.

For example, the polynomial $X^3 + X^1 + X^0$ (also written $X^3 + X + 1$) represents the bit location of $2^3 + 2^1 + 2^0$ in binary, which is 8 + 2 + 1 in decimal, which is the prime number 11. The LFSR that represents this prime polynomial is shown as the PRPG in Figure 3-27. If this PRPG LFSR is initialized with the all 1s seed and a clock is applied repeatedly, then the result will be the following sequence of 3-bit values:

111 --> 011 --> 001 --> 100 --> 010 --> 101 --> 110 --> 111

As can be seen, all 2^n -1 (2^3 -1), or 7 states, are created.

3.27.5 X-Management

Another pitfall to LBIST is that the result captured in the signature register on each clock cycle must be absolutely deterministic. If there is one legal X value (an allowed ambiguous state), then there are two valid signatures, if there are two X values, then there are four valid signatures, 3 Xs equal 8 signatures, and so on. It gets out of hand fast. This means that the circuit under test must meet more DFT rules than just those applied for full-scan compliance. There can be no partial scan (uninitialized states), no multiple cycle timing paths (transparent latches, tristate nets, long paths), and no non-deterministic logic at any location that the signature analyzer captures data.

3.27.6 Aliasing

Another concern with using signature analyzers has to do with the concept of more than one fault resulting in a self-repairing signature. The possibility that this could happen is known as aliasing. Aliasing is usually described as a probability of two faults resulting in a correct signature. Since the signature analyzer is just a register that can be viewed as a counter-like device (with each state possible in the signature register being a count value), then the probability is 1 in however large a count or 1 in 2^n (usually written as $1/2^n$ or 2^{-n}).

For example, if an LFSR was 4 bits long, then it would have 16 possible state values (2^4, 0 to 15 in decimal). If the input to this LFSR was connected to a circuit with 4 outputs, then it would have a count value associated with each data input at each clock cycle. If exactly 16 clock cycles were applied, and each clock cycle mapped to a different count, then aliasing would be the probability that two or more errors would map to the count value that represents the good signature. In this case, the probability is 1 out of 16, or as it commonly presented, $1/2^4$.

However, aliasing-like behavior is not just limited to multiple failures. It has implications for fault diagnosis as well. If more than 16 clock cycles were applied, then a repeat of one of the state values would occur. For each clock cycle past 16, then the signature analyzer can repeat the various state values. So several configurations of good and faulty behavior may map to the same count values (for example, a perfectly good circuit that requires 32 clocks to test fully may pass through all 15 error signatures twice, and may have the passing signature represented twice). The incoming data to the signature analyzer also directs the analyzer to the various states. It may be possible to provide an input stream that would cause the signature analyzer to represent only 3 or 4 of its count states repeatedly. So the selection of a signature analyzer is important if it is to be trusted for its final signature or if it will be used for any sort of diagnostic development. Diagnostic development is generally done by keeping a signature history during simulation or a signature dictionary as a look up table (known single faults result in known signatures).

One other problem with a signature analyzer is the "zero-ing out" problem. This is another version of the loss of diagnostic information due to the repeating of the state values of the LFSR. In this case, the repeated state is the zero (or some other) value which can be viewed as the erasing of the LFSR's history. For example, if a circuit with a fault responds to applied input values with a data word that is the natural polynomial value, then the signature analyzer will "zero out"—return to its seed value (note: the signature analyzer LFSR can have the all zero state; the PRPG LFSR cannot). If the signature analyzer keeps "zero-ing out," then the history of the circuit response is lost and the resultant signature has minimal value (only a few cycles of circuitry response = history).

The possibility of aliasing, overcount, and "zero-ing out" decreases with the length of the LFSR, since it is defined as $1/2^n$, so a solution to minimize these problems is to make the signature analyzer longer (wider) than the response word—for example, using a 20-bit LFSR for a 16-bit data bus. This increases the count values from 2^{16} to 2^{20} and therefore provides more possible signature values (1,048,576 as opposed to 65,536).

Scan Testing Methodology

Advantages

Direct Observability of Internal Nodes
Direct Controllability of Internal Nodes
Enables Combinational ATPG
More Efficient Vectors
Higher Potential Fault Coverage
Deterministic Quality Metric
Efficient Diagnostic Capability
AC and DC Compliance

Concerns

Safe Shifting
Safe Sampling
Power Consumption
Clock Skew
Design Rule Impact on Budgets

Figure 3-28 Scan Test Fundamentals Summary

3.28 Scan Test Fundamentals Summary

This chapter has covered the fundamentals of scan and some of the scan techniques and scan test architecture components used in the industry today (se Figure 3-28). The key items of information on scan, scan architectures, scan techniques, and the scan methodology that should have been learned in this chapter are:

- Scan is a structured design-for-test methodology and is automatable and repeatable.
- Scan allows the direct control and observation of internal circuit nodes via scan chains made of scan cells, which are scannable flip-flops.
- Scan is designed predominantly for structural test.
- Scan enables ATPG and the ATPG methodology to be used—and full-scan reduces the problem to an effective combinational-only problem.
- Scan vectors are more efficient than functional vectors, are generally associated with a higher fault coverage metric, and provide a deterministic fault metric when generated.
- Scan architectures with multiple balanced scan chains result in very efficient vector sets that are more likely to reduce the cost-of-test.
- Scan enables testing for both AC (timing) and DC (structure) test goals.
- Scan can enable a very efficient, automatable diagnostic capability.

However, not everything is totally positive about the scan methodology. The concerns with scan, which requires concurrent design/test engineering tasks to be done are:

- Safe scan shifting requires additional control logic and design analysis.
- Safe (contention-free) sampling requires additional control logic and design analysis.
- The application of design rules to meet tester, ATPG, or safe scan requirements may affect design budgets and requires test-related tasks to be scheduled during the design phase.
- Scan requires more concern about the clock-delay and clock-skew parameters of the physical design.
- Scan may require bypass clock inputs on chip designs that contain on-chip clocks—multiple time domain chips may require multiple bypass clocks and multiple scan domains, each with its own SE control signals.
- Scan may target operation at a higher speed than the tester's clock or data rate.
- Critical path selection for AC path delay scan-based testing is necessary to reduce the amount of vectors that may be delivered.
- Compressed scan vectors may reduce the size of the test data, but they are more active and may elevate the chip's power consumption during test.

3.29 Recommended Reading

To learn more about the topics in this chapter, the following reading is recommended:

Abromovici, Miron, Melvin A. Breuer, Arthur D. Friedman. *Digital Systems Testing and Testable Design.* New York: Computer Science Press, 1990.

Dillinger, Thomas E. *VLSI Engineering.* Englewood Cliffs, NJ: Prentice Hall, 1988.

Gulati, Ravi K. and Charles F. Hawkins, eds. *I_{DDQ} Testing of VLSI Circuits—A Special Issue of Journal of Electronic Testing: Theory and Applications.* Norwell, MA: Kluwer Academic Press, 1995.

IEEE Standard Tests Access Port and Boundary-Scan Architecture. IEEE Standard 1149.1 1990. New York: IEEE Standards Board, 1990.

Parker, Kenneth P. *The Boundary-Scan Handbook, 2nd Edition, Analog and Digital.* Norwell, MA: Kluwer Academic Publishers, 1998

Rajski, Janusz, and Jerzy Tyszer. *Arithmetic Built-In Self-Test For Embedded Systems.* Upper Saddle River, NJ: Prentice Hall, 1998.

Tsui, Frank F. *LSI/VLSI Testability Design.* New York: McGraw-Hill, 1987.

CHAPTER 4

Memory Test Architectures and Techniques

About This Chapter

This chapter contains the Memory Test Architectures and Memory Design-for-Test Techniques section that has been designed to teach the basic and practical fundamentals of memory test and memory design-for-test. This chapter is not an exhaustive comprehensive text on the minutia involved with memory failure mode analysis and the multiplicity of memory test algorithms that have been developed to assess defect coverage to the nth mathematical degree, but is instead a basic treatise on the interaction of the memory design concerns with memory test concerns and on how these concerns are generally handled in the industry.

This chapter includes material on memory design and integration concerns, memory test concerns, types of memory, basic memory testing, common memory fault models, common memory test methods, common memory test algorithms, the interaction with memory and a scan test architecture, and the use of memory built-in self-test.

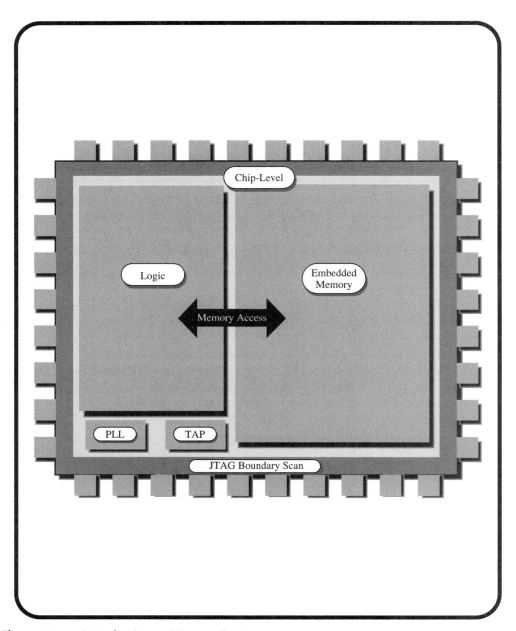

Figure 4-1 Introduction to Memory Testing

4.1 Introduction to Memory Testing

4.1.1 Purpose

The Memory Test chapter describes the basic components of memory testing, and then describes the test techniques and architectures that are commonly used in many parts of the industry today. These memory test techniques and architectures described allow memory test architectures to co-exist with scan and other test architectures to produce an efficient overall chip-level test architecture, and/or will allow many memory arrays to be tested simultaneously by the use of embedded memory built-in self-test (MBIST). This information is provided to clarify the trade-offs involved with the various memory test processes so that the correct techniques and architectures can be applied to meet the cost-of-test, time-to-market, and quality-level goals of a semiconductor provider.

4.1.2 Introduction to Memory Test

The leading indicator of a semiconductor manufacturer's product and process technology and quality can be found in their memory array technology. For most manufacturers, the memory array is the most dense physical structure and is made from the smallest geometry process features available. In today's market, the embedded memory array structures are beginning to dominate the physical die area versus the die area dedicated to logic (see Figure 4-1). Therefore, the memory arrays are more sensitive to defects (there is a higher probability that the random defect content will land on and will have a deleterious effect on the memory arrays rather than the general chip logic).

The current semiconductor market is driving memory technology to become a larger test problem. The memory is already one of the most sensitive areas for defect landings, and now the industry is supporting even greater amounts of memory being embedded into chip designs without sufficient direct access to the chip pins (and therefore, no easy access from the tester). Some memory arrays are even being "doubly embedded" in that they are designed into a hard macro-embedded core device that is then wholly embedded within a chip design (for example, embedding a microprocessor with a cache within a larger chip design)—if insufficient test access exists the first time the memory is embedded, then the problem only gets worse as the memory is layered deeper within a chip design.

The structured memory is already a difficult device to test, because it is a structure that is unique and in a way "custom," since it can't be easily modeled with gate-level equivalents (representing a memory array with 8192 words that are 32 bits wide [a 32x8K] with gate-level equivalents requires 262,144 flip-flops and a large amount of combinational decode logic). The memory test is further complicated, since it is also a "defect" oriented test. Memory testing has evolved into defect testing, since the "failure mode and effects analysis" (FMEA) of defects versus memory operation are well understood, and because of the sensitivity of the memory to defects resulting from the transistor density. The structured nature of the memory, the inherent controllability and observability at the array level, and the requirement of defect-based testing have resulted in the creation of "algorithmic testing" and reliance on "multiple clue" analysis to detect and identify the source of any failures.

In the modern chip design, or system-on-a-chip design, the topic of memory testing is not limited to just testing the embedded memory by using a tester. The embedded memory and its related test support must also co-exist with the test structures used for logic. So the memory, which may be tested by direct tester access or by some form of built-in self-test structure, must also not interfere with any logic testing structures such as scan design. If a built-in self-test methodology is used, then some or all of the address, data, read/write, and memory test control logic will be included within the chip, so there is the extra step of considering the extra test logic as part of the design, and this test logic must meet the same engineering design budgets and constraints applied to all other chip logic (power, area, frequency, etc.).

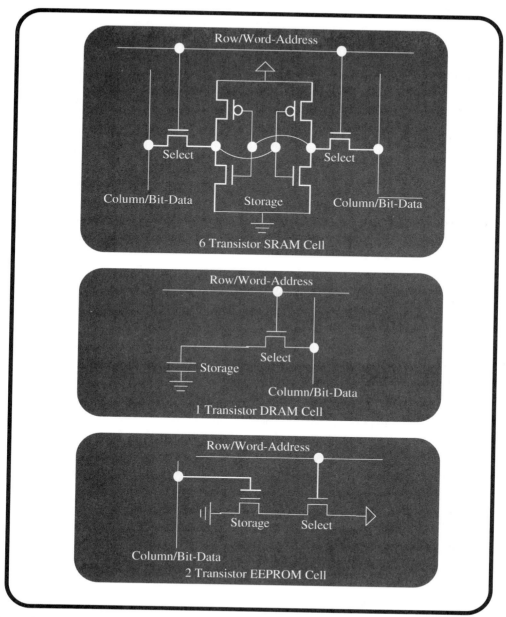

Figure 4-2   Memory Types

4.2 Types of Memories

4.2.1 Categorizing Memory Types

Many types of memory circuits are used in the industry today, and the individual memory array circuit configurations can vary greatly. For embedded CMOS (Complementary Metal Oxide Semiconductor) memory arrays, the most common volatile structured memories are: static random access memory and dynamic random access memory, SRAM and DRAM, respectively. Volatile memory arrays require that power be applied continuously or else the memory will lose its contents.

Another type of memory that is common in the industry is the non-volatile memories—memories that keep their content even when power is removed. The most common representatives of these types of memory are electrically programmable read-only memories (EPROM)—for example, EEPROM and Flash—and ROM. Each of these have different physical circuit configurations, and so have different failure modes, testing requirements, and cost-of-test trade-offs (see Figure 4-2).

SRAM The most common embedded memory type used for embedded RAMs, Data Caches, and Tag arrays is the Static Random Access Memory (SRAM). SRAMs can, theoretically, hold the data in the memory cell indefinitely, or until a write is accomplished, as long as DC power is applied to the chip. The memory is considered volatile, since the removal of DC power will result in the loss of memory contents. Only a write cycle will affect the memory cell contents—this means that a read cycle will not affect the memory cell contents. The SRAM is considered an "expensive" memory structure since generally six transistors make a single memory cell (a bit of data).

DRAM A more area efficient memory is the Dynamic Random Access Memory (DRAM). More memory can be embedded, since the storage cell has fewer transistors, but a more complicated fabrication process is required to create the capacitor that is used for data storage. The DRAM also has an AC power refresh requirement, since storage is handled by a capacitor and the capacitive charge can bleed off over time. The configuration of this memory allows a read operation to destroy the memory content (destructive reads). This type of memory is also volatile in that the memory loses its data when power is removed from the chip.

EEPROM EPROM is a class of memory arrays that allows the contents to be programmed, and the data contents remain after the chip's power is removed. These types of memory arrays are referred to as non-volatile. One of the most common types of these memories is the Electrically-Erasable Programmable Read-Only Memory (EEPROM). The EEPROM and a related memory type, Flash, also require a more complicated process to implement, since the storage device relies on a floating gate and charge transfer (Flash is a type of programmable memory that is block or page addressable, whereas EEPROM supports the more traditional word addressing scheme). EEPROM allows data to be written into the memory when a higher than normal voltage is applied (for example, 7 volts within a 3.3 volt CMOS chip).

ROM Another memory type that allows the memory data to remain when the chip's power is removed is the Read-Only Memory. The most common embedded memory in this class is the "Mask Programmable" ROM, in which the memory array contents are installed during the manufacturing process. Note that once the data is installed within the ROM memory, it can never be altered without re-processing the silicon.

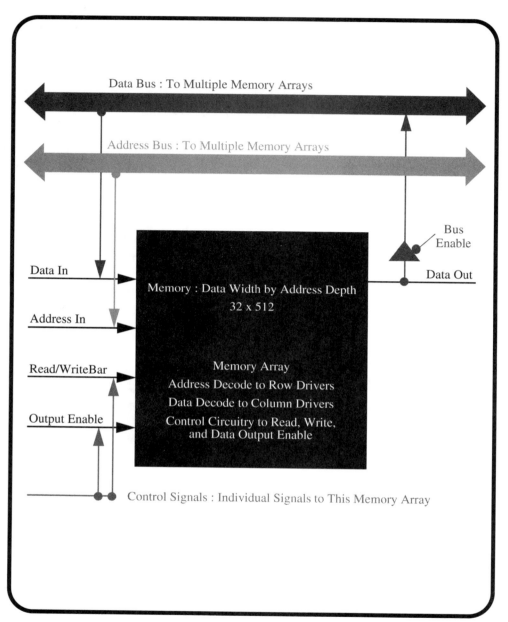

Figure 4-3   Simple Memory Organization

4.3 Memory Organization

4.3.1 Types of Memory Organization

When memory arrays are embedded within a chip, the memory data and control signals may or may not be accessible by the chip's package pins as part of the functional design. In general, the embedded memory arrays are connected to internal chip-level data and address busses that also service the embedded processor circuitry. There are many ways to decode the memory arrays within the processor's address space, from continuous and contiguous memory mapped space to banked or paged memory mapped space.

The key information to be noted here is that individual memory arrays are connected to data busses, address busses, and control circuitry that can request a read or write operation or enable the memory output to the data bus (see Figure 4-3). If several separate memory arrays are supported on the bus structure, then they may share the data bus and different decodings in address space; note that not all memory arrays, attached to any given data bus, have the same data width and address depth requirements.

4.3.1.1 Address Decoding

The address selection signals can be dedicated signals from a logic source routed to a specific memory, or they can be delivered as a set of multiple-driver tristate nets (an address bus). For any given memory, the address lines are used to select which word in the memory will be read or written. The number of address signals required is N, where N is from the nearest 2^N that encompasses the entire number of words in the memory array. For example, a memory that has 32-bit-wide data words and 256 words would have $2^N >= 256$ or N=8. Note: another way of viewing this is to use Log_{10} math where a general equation is "Address lines" = (Log_{10} "Number of Words")/($Log_{10}2$).

The system address lines are decoded and passed into the memory array as row decoding.

4.3.1.2 Data Decoding

The data signals can be dedicated signals from a logic source routed to a specific memory, or the data signals can be delivered as a set of multiple-driver tristate nets (a data bus). The data lines are used to provide the logic state associated with each bit within a word.

The data bus lines are decoded and passed into the memory array as column decoding.

4.3.1.3 Control Signals

The type and number of control signals may differ with different memory array types. Some memory arrays have a single signal that represents the "Read-WriteBar" signal to control whether the memory is conducting a read or a write; some memory arrays may have separate signals to indicate a "read" versus a "write;" and other memory arrays may have separate "read," "write," and "data output" enable signals. Depending on whether the memory array is synchronous or not, these signals may directly control the operation (in a combinational sense), or these signals may be used in conjunction with a synchronizing clock. When multiple control signals are supported, the memory array may support the concept of a "stall" or "idle" (the memory does not need to commit an action or operation every cycle; it can just sit and do nothing).

The described control signals are mostly applicable to SRAM and DRAM types of memories, ROM, EEPROM, and Flash memory arrays do not have traditional "write" control logic. The ROM, being read-only, has only read control signals, and the programmable memory, EEPROM and Flash, may have only "program" and/or "erase" control logic.

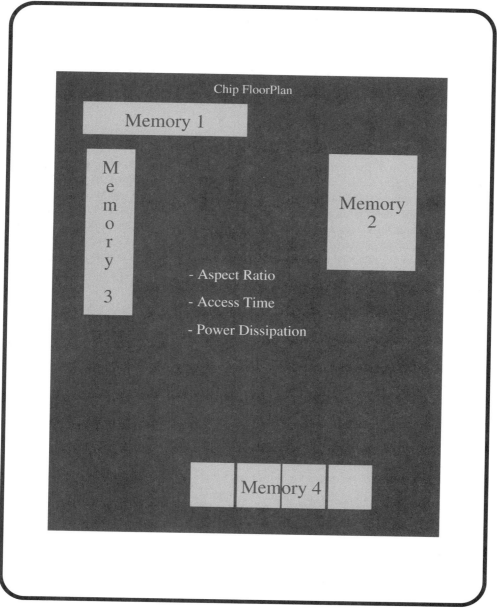

Figure 4-4 Memory Design Concerns

4.4 Memory Design Concerns

4.4.1 Trade-Offs in Memory Design

When memory arrays are embedded within a chip, physical design concerns are associated with each memory. For example: Is the memory going to be square, rectangular, or broken into several blocks? Is the access time performance going to be within some specification? And what is the power dissipation of the memory? These three design issues are all interrelated and affect each other (see Figure 4-4).

4.4.1.1 Aspect Ratio

The relationship of the memory array's X-length to Y-height is known as the aspect ratio of the memory. The aspect ratio is a design concern, because it is a trade-off in placing the memory on the die in such a way that it does not impact the area or the ability to successfully route the chip. Placing the memory as a square, vertical rectangle, or horizontal rectangle, or breaking the memory into smaller pieces are all methods that can be used to meet chip area goals. However, the aspect ratio, or breaking the memory into several smaller blocks, may affect the performance of the memory array, since the row or column drive signals may now be longer or shorter or replicated multiple times, depending on the layout decisions. The length and number of the row and column driver routes can affect power usage/dissipation.

4.4.1.2 Access Time

Another memory design concern is the access time or time performance of the memory. If the memory must be made to respond to read or write requests within some time specification and a long-skinny aspect ratio is also stipulated, then the row or column driver logic may need to be excessively upsized in drive strength to meet the performance goals (this increase may negatively affect area and power dissipation).

4.4.1.3 Power Dissipation

Long row and column drive signal routes or multiple copies of the row and column drivers can increase the power requirements of the memory array. Upsizing the drivers to meet an aggressive access time also increases the power requirements.

Generally, power ratings are described as functions of aspect ratio, access time, and operation (read, write, stall). Power ratings are usually specified as milliwatts per MegaHertz per operation (e.g., 35 milliwatts per MegaHertz for a read—if the memory is to be operated at 20MHz, then the rating is 700 milliwatts or .7 watts).

4.4.1.4 Special Cases

The "write" or "programming" time of EEPROM or Flash is much longer than a "write" cycle in DRAM or SRAM. For this reason, test time can be adversely affected (extended into the several second range). For example, Flash can be only block "programmed" to a logic 1, and must be block "erased" to a logic 0—the program and erase cycles can be hundreds of milliseconds in duration. Normal read-write-read operations, if used for test purposes, would not be "cost-of-test" feasible.

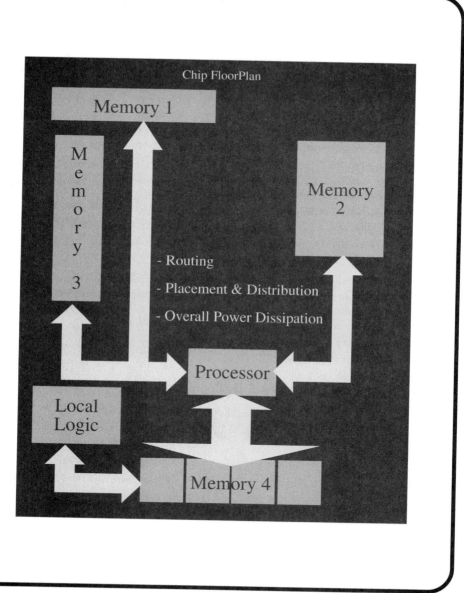

Figure 4-5   Memory Integration Concerns

4.5 Memory Integration Concerns

4.5.1 Key Issues in Memory Integration

When memory arrays are embedded within a chip, there are physical design concerns at the chip level for the sum total of memories. For example: How many memories will be integrated? What are the aspect ratios of all of the memories? Where are they to be placed? What is the routing impact for this number of memories and distribution? And what is the power dissipation of the sum total of all the memories? These design issues are all interrelated and affect each other (see Figure 4-5).

4.5.1.1 Number, Placement, and Distribution of Memories

A more recent trend in chip design is to have many distributed memory arrays rather than just one contiguous memory or just a few (for example, 30 memory arrays versus 3 memory arrays). Shrinking process geometries and increasing integration levels has been the driver for this trend. With more embedded memories, some chip-level decisions must be made on the number and physical floorplanned location of the memories: whether or not their address selection is from some common control point, or whether their address selection is accomplished individually by local circuitry.

Historically, memory has been placed in one area of the chip (for example, such as on one side of the die) and has been accessed from a common control point or address/data bus. Shrinking process geometries have allowed more and more memory to be integrated on-chip. The trend has not been to keep the memory contiguous, but to distribute it across the chip near the logic that the memory services (and this supports better performance by reducing routing requirements). This means that the aspect ratio of each memory, the breaking of memory arrays into smaller blocks, the number of memory arrays, and their placement with respect to each other are tied to the chip-level goals of area, power, and the timing performance of each memory.

4.5.1.2 Chip-Level Routing Impact

The shrinking geometries that allow more memory integration also poses another problem—routing delay. In deep-submicron design, routing delay is a major concern and may impact the access time or time performance of the overall memory architecture. If the memory is placed as a contiguous structure, then megabytes of memory might be accessed by a single address-data bus structure. If the memory is to be accessed from several different functional areas, then a wire-intensive data-address structure may have to be routed from several locations across the chip to the memory structure. However, if the memory architecture is widely distributed, then the routing (and delay) impact may only be localized.

4.5.1.3 Chip-Level Power Dissipation

At the chip level, the overall memory architecture may be limited by the sum of the power consumption of all memory arrays, the chip's power structure, or the chip's package. If all memories can conduct a worst-case operation (e.g., read, write, or stall) on a single clock cycle at the maximum rated frequency, then the chip power structure and the package should be designed to handle this power rating.

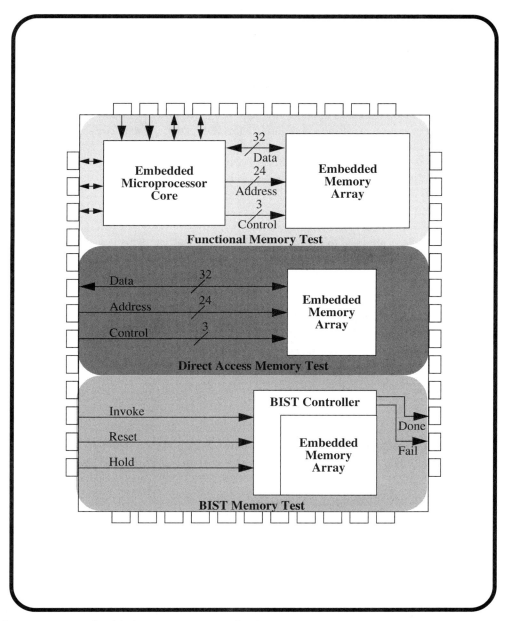

Figure 4-6 Embedded Memory Test Methods

4.6 Embedded Memory Testing Methods

4.6.1 Memory Test Methods and Options

When memory arrays are embedded within a chip design and there is no direct "functional" access to the memory then a testing problem exists (some designs support the direct access of the embedded memory arrays through dedicated functional pins—these cases generally do not pose a controllability/observability test problem). Three common or basic test methods can be applied, and each of these methods has trade-offs that involve: cost-of-test, chip area, chip pin requirements, chip-level timing, and to some extent, chip power requirements. The three test method options are: Embedded Microprocessor Access (see Figure 4-6); Direct Memory Access (Ad Hoc or Test DMA); and Built-in Self-Test (MBIST).

4.6.1.1 Embedded Microprocessor Access

The first kind of testing that a designer usually thinks of is, "Let's just use the embedded microprocessor that is attached to the memory to test the memory." This is viable in certain circumstances. The trade-offs for this type of memory testing option are:

- The vectors are actually microprocessor code (compiled assembly language), applied at the full chip interface, which operates the memory by way of the microprocessor, and these vectors must reside in (or consume) tester pattern memory.

- The vectors must be developed manually as opposed to ATPG or a tester-based ATG option.

- The pattern sequences applied to the memory will be functional or operational patterns only, and the sequences and allowed operations may not be able to exercise all structural failure modes.

- Failure information may be provided as chip-level microprocessor outputs that must be analyzed, processed, or filtered—further debug and analysis work must be done beyond just the first level of data collection.

- Note: if the microprocessor and the memory array are bundled into an embedded core unit, then "direct functional access" may not exist in the final integration—the operational vector development then becomes a very complex problem and vectors must be developed late in the development cycle, possibly after the delivery of silicon.

4.6.1.2 Direct Memory Access

Another test method option is to provide a test architecture that allows direct access to the embedded memory arrays. The access can be from the package pins directly to an embedded memory, or it can be from package pins to some test conditioning logic and then to some pipelined sequential stages that feed the memory. The trade-offs with this type of memory testing option are:

4 • Memory Test Architectures and Techniques

- Signal pins, of sufficient quantity to provide a data, address, and control interface must be supplied or borrowed (e.g., it depends on the size of the memory, but is usually about 60 pins) and modified to support direct memory testing.
- A route-intensive bus structure must be designed for and delivered to each embedded memory (the problem gets worse as more memories are supported—a change in any memory structure requires test architecture design adjustment).
- Vector generation can be done with a "Memory Test Function" (MTF) provided by some ATE. These MTFs do not use pattern memory—and these options in many cases may also provide failure "bitmap" capability.

4.6.1.3   Memory Built-In Self-Test

A more recent method of memory testing is to include the tester function right in the silicon, by placing an address generator, data generator, and algorithmic sequencer into the silicon design. The trade-offs involved with this type of memory testing architecture are:

- Only a small quantity of chip-level interface signals must be borrowed to provide at least the "BIST-invoke," "BIST-reset," "BIST-done," and "BIST-fail" chip-level test interface signals.
- The BIST test algorithm, once selected and designed, is fixed in the silicon and can't be changed (unless a very complicated programmable BIST is designed)—and the diagnostic bitmap may be more difficult to build if access to fail data is not adequately provided.
- The memory BIST can operate the memory at speed if necessary with just the application of a clock signal, or by relying on any internally generated clock (such as a PLL). A high-speed memory can be fully verified by a tester that can't support the chip's data rate.
- The vector to operate BIST is the transitioning of the "Invoke" signal and then the application of clocks until a "Done" or "Fail" signal indicates that the test has completed. A clocking loop does not consume vector space in the tester.
- A chip-level BIST test architecture can be applied to multiple memories with the option of one controller and routing to several memory arrays, or as multiple BIST controllers associated closely with each memory array.

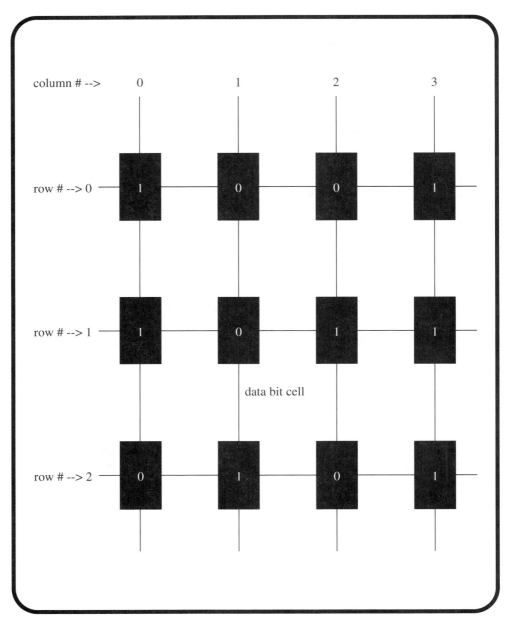

Figure 4-7 Simple Memory Model

4.7 The Basic Memory Testing Model

4.7.1 Memory Testing

Memory array circuitry is handled differently from general chip logic in both design and test. Memory arrays are much more dense than general chip logic because memories are very regular structures. This fact makes memory arrays more susceptible to silicon defects.

The regularity of memory logic (see Figure 4-7) is also an advantage in that it allows memories to be tested directly for defect models, instead of just fault models, by using inductive analysis. Inductive analysis is multiple clue analysis (similar to the way Sherlock Holmes solves a case). This testing is different from general logic testing, which relies on the single fault assumption.

4.7.2 Memory Test Fault Model

The basic memory fault model is that a memory cell, or bit, does not have the correct data at the observe (read) time. This basic fault model covers all the complicated memory failure mode actions. In other words, for fault detection, it doesn't matter how the memory cell gets the wrong data; the key is that it has the wrong data. However, for diagnosis, the mechanism of the failure is important. This means that memory testing is accomplished by placing data within a memory cell, and then reading that cell and verifying that the data has not been corrupted (e.g., writing to the memory and then reading the written data at some later time).

4.7.3 Memory Test Failure Modes

The whole point behind memory testing is to apply several different data values and data value sequences to the memory array and to discern the failure mode (cause of the failure) by identifying which tests (sequences) fail and which do not.

The failure modes that can result in wrong data are (mode:definition):

- Data Storage: that the bit cells accept, hold, and return data
- Data Retention: that the bit cell data does not "leak" or degrade over time
- Data Delivery: that the sense lines can apply read/write operations
- Data Decode: that the proper data is delivered to the correct sense lines
- Data Recovery: verification of the access time and sense-line recovery time
- Address Delivery: that the row drivers can apply address selection
- Address Recovery: verification of the address decode/selection time
- Address Decode: that every word can be selected—and uniquely
- Bridging: that one part of the memory does not disturb another part
- Linking: that one part of the memory does not freeze another part

The different types of failure modes, mechanisms, and effects will be described further in the following sections.

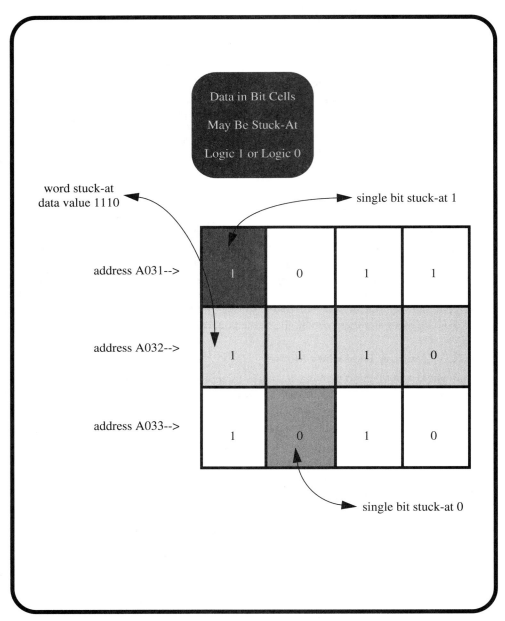

Figure 4-8 Bit-Cell and Array Stuck-At Faults

4.8 The Stuck-At Bit-Cell Based Fault Models

4.8.1 Stuck-At Based Memory Bit-Cell Fault Models

Just as with general logic testing, a memory bit cell can be modeled as a stuck-at fault (see Figure 4-8). Since memory arrays are regular structures, the stuck-at model may be applied to any portion of memory, such as: a bit, nibble, byte, word, longword, or block.

The application of the stuck-at model is to "stick" the bit data value of an individual cell or group of cells to some constant logic value. For individual cells, the value may be a logic 1 or a logic 0. For a larger portion of memory, the value may be all logic 1s, all logic 0s, or some data value (for example, the byte value 0110).

4.8.2 Stuck-At Fault Exercising and Detection

The source of the Bit-Cell stuck-at fault may be a defect in the bit cell power, ground, select, or storage transistors. The stuck-at is detected by writing both a logic 1 and a logic 0 data value to every data bit cell and by reading the respective logic 1 and a logic 0 from every data bit cell.

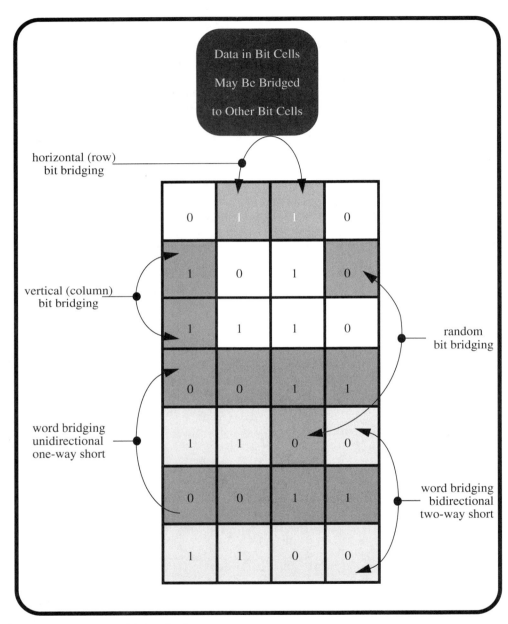

Figure 4-9 Array Bridging Faults

4.9 The Bridging Defect-Based Fault Models

4.9.1 Bridging Defect-Based Memory Test Fault Models

One of the most common concerns with memory arrays is the bridging fault (see Figure 4-9). The high density and closeness of the memory array elements lead to a higher probability of physical bridges between adjoining cells. One of the oddities of memory design is that adjoining cells are not always in logical order (i.e., in a contiguous word or address space order). This fact makes the bridging harder to detect, since any two physical adjoining cells may not be related logically at all (e.g., bit-4, word-3 may be physically placed next to bit-8, word-256 if the memory array is made of 4x4 tiles).

Bridging faults may be 0 Ohm, resistive, or diodic physical bridges. The 0 Ohm bridge may cause two bit cells to follow each other (a change in one results in the change in the other). The resistive bridge may cause a delay fault (the cell is slow to change from one logic state to the other). The diodic bridge can look like an intermittent fault, since only one memory element (bit, nibble, word, etc.) will change value and only when the memory element at the source end of the bridge changes (and only when the changing source data is different from the destinations target data). A bidirectional diodic bridge looks very much like a 0 Ohm short. This "scrambling" and the many types of bridges make the bridging test a very complex test.

4.9.2 Linking Defect Memory Test Fault Models

The linking fault is very similar to the bridging fault, except that a change in data in some bit cell or address location causes the data in some other bit cell or address location to freeze (locked up so that the data can't be changed at a later time). In reality, this fault effect seems to be an intermittent stuck-at fault—the location may pass and may pass several times, and then start failing after the linking location is accessed, and only after different data is expected from the linked location.

4.9.3 Bridging Fault Exercising and Detection

The bridging fault is detected by writing alternating or complementary patterns that would place opposite data in physically adjoining cells. If the physical placement of the cells is not known to the test developer, then general complementary background patterns (0-F, 3-C, 5-A, etc.) should be applied that would exercise all possible bridges (bits, nibbles, bytes, words, longwords, etc.).

Also note that the unidirectional diodic bridge may be sensitive to the address order. If data is written and read from address 0 to address N (N being the max address value such as 255 for an 8-bit by 256-word memory), then a unidirectional bridge may destroy data in a lower address when a higher address is written. For example, writing data into the word at address 128 may corrupt data in the word at address 3, and if the applied test is only from address 0 to address 255, then the corruption will not be discovered during the test process. For this reason, the bridging patterns must be applied in both forward and reverse address order—and multiple times, since the first access may "exercise" or "excite" the fault, but only multiple accesses will detect the fault.

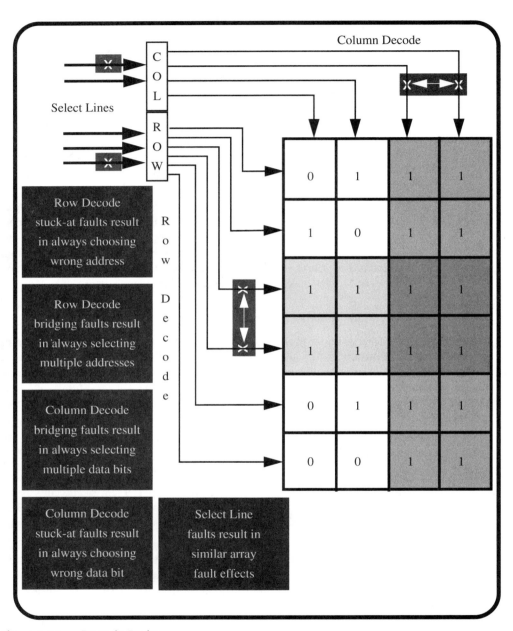

Figure 4-10 Decode Faults

4.10 The Decode Fault Model

4.10.1 Memory Decode Fault Models

The memory decode fault can occur at any of several levels (see Figure 4-10). The fault model can be applied at the local memory address and data busses, at the system level address and data busses, or at the intra-memory decoding that involves the internal array row and column drivers and the read-write control logic. The fault models that can be applied are the stuck-at, stuck-open, bridging, and delay fault models.

4.10.1.1 Data Bus and Column Decode Faults

Fault models applied to the data bus signals and to the column driver decode and to the column drivers represent bits within a word. These fault models result in bits being stuck at a logic value throughout all or some portion of the address space (for example, a single column may control the data values in bit 27 of the whole address space of a 32-bit wide memory); or two data bits may be shorted together, or the access time for one data line may be extended.

4.10.1.2 Address Bus, Address Decode, and Row Decode Faults

Fault models applied to the address bus, address decode, row decode, and the row drivers represent the address space of the words in a memory. These fault models result in the selection of the same word/address, regardless of the applied address (for example, a single row may represent word 96 out of 128 words in a memory array), or two words may be selected simultaneously, or the access time for one or more words may be extended.

4.10.2 Decode Fault Exercising and Detection

The types of fault models that may exist are stuck-ats, bridges, and delay faults due to shorts, opens, bridges, and other process errors. The patterns needed to exercise and detect stuck-at decode faults are those patterns that would apply both logic 0 and logic 1 to every data and address line and in such a way that each address can be resolved uniquely (if the entire memory had only the data value A, then the selection of the wrong address would go undetected). Delay faults and bridges can be exercised and detected by applying logic transition (vector pair) vectors to the address space and the data bit lines. Bridges are also tested similarly to data cell bridges, but the fault model only needs to be applied to the address and data lines and decode logic (the physical layout of the memory is not required, since the decode represents the logical organization of the memory).

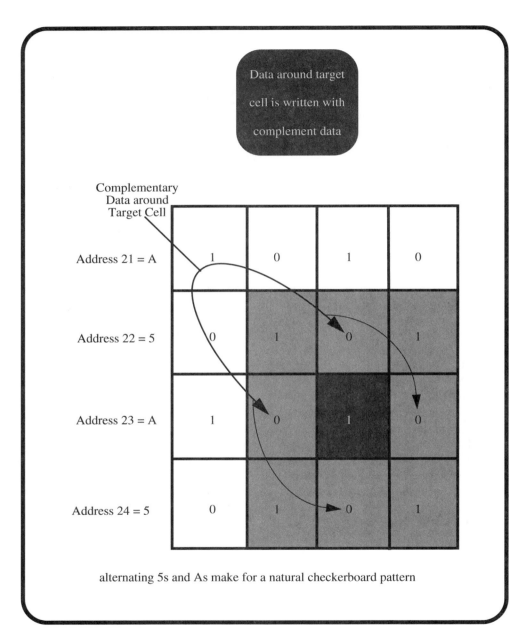

Figure 4-11 Data Retention Faults

4.11 The Data Retention Fault

4.11.1 Memory Test Data Retention Fault Models

One of the concerns with memory arrays is that leakage or bridging may cause the stored data value to degrade over time. The normal method for verifying data retention is to put data in the memory, leave it there for a period of time, and then to read/verify the data. To exacerbate the conditions of the test, a forced leakage state can be emulated by placing a "surround by complement" or "checkerboard" pattern in the memory array (see Figure 4-11), or by modulating the voltage/temperature levels during the testing process. Once the data is placed in the memory, then the clock may be stopped and a static pause may be accomplished. Another method can be to place data in the memory, to then "write lock" the memory, to continue testing other logics, and return to read and evaluate the memory at some later time.

4.11.2 DRAM Refresh Requirements

The DRAM memory array has a capacitor as a storage element. Since a capacitor has a known time constant associated with the storage, then the data must be re-written or refreshed as a normal functional operation. This increases the test complexity of the DRAM in that the testing sequence must also allow the refresh cycling, and the refresh cycle is now an operation that needs to be verified as part of the test process. The refresh, for example, would prohibit continually reading and writing from the top of the memory, to the bottom of the memory if this process would exceed the refresh period.

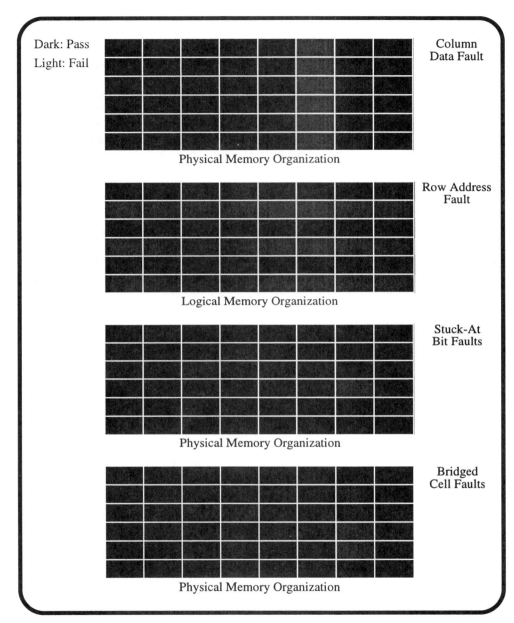

Figure 4-12 Memory Bit Mapping

4.12 Diagnostic Bit Mapping

4.12.1 Memory Test Diagnostics: Bit Mapping

As was described in this chapter so far, many types of pattern sequences need to be applied to detect and identify failed bits. The failures may occur due to stuck-at data cells; delayed access time from certain data cells; stuck-ats, bridges, or delays related to address, data, row, and column decode; and corrupted data values due to leakage or inadequate refresh. If all the failures (incorrect data at the observation of a read), regardless of cause or the applied test, are mapped on an X-Y graph of the physical cell structure of the memory layout, then the result is a physical failure bit map. If the same failures are mapped onto a graph of the logical cell structure of the memory layout, then the result is a logical failure bit map (see Figure 4-12).

When the bit maps are examined visually, sometimes organized patterns appear. These patterns are the key to "multiple clue" analysis or "inductive" testing (deducing the failure mechanism by evaluating all the clues). For example, if several patterns are written into the memory, and every failure is logged and mapped, and the resultant logical map shows that every other or every fourth word is failing in its entirety, *then* the suspected failure mechanism is an address line stuck-at.

Other common patterns are:

- A vertical stripe can be a column driver or data bit-line stuck-at
- A horizontal stripe can be a row driver stuck-at
- Groups of vertical stripes can be data decode, data bit-row driver bridges
- Groups of horizontal stripes can be address decode, column-address bridges
- Individual non-related failures may be multiple data bit cell stuck-ats
- Paired failure groupings on the physical mapping may indicate bit cell bridges

As can be seen, the generation of a bit map can accelerate the debug and diagnostic process by quickly pointing to "most likely" defect sources.

Figure 4-13 Algorithmic Test Generation

4.13 Algorithmic Test Generation

4.13.1 Introduction to Algorithmic Test Generation

As was described in each failure mode section, certain types of pattern sequences need to be applied to exercise and detect the different failure modes. Instead of applying separate patterns designed to aggravate each of the fault classes, a great amount of research has gone into defining pattern sequences that exercise and detect as many fault classes in the least number of operations (conserving test time) and into defining pattern sequences that detect the hard-to-get failure modes (see Figure 4-13).

For example, to exercise and detect stuck-at faults, bridging faults, address decode faults, data decode faults, and access time faults, a March algorithm such as the 14N March C+ type test could be used (where 14N is the number of operations per memory word). A March C+ type algorithm, for example, is described as follows (in shorthand notation—W=write, R=Read):

- Address (0) --> Address (Max): W(5)
- Address (0) --> Address (Max): R(5)-W(A)-R(A): Increment (Address)
- Address (0) --> Address (Max): R(A)-W(5)-R(5): Increment (Address)
- Address (Max) --> Address (0): R(5)-W(A)-R(A): Decrement (Address)
- Address (Max) --> Address (0): R(A)-W(5)-R(5): Decrement (Address)
- Address (Max) --> Address (0): R(5)

Note that the initialization is 1 operation, the RWR is 3 operations times 4 repeats, and the final read verification is 1 operation—this total is 14 operations per word or 14N. This test will write every location with a logic 1 and a logic 0 (getting the stuck-at faults); will conduct read-write-read at speed (getting row and column driver recovery and delay faults); will write to addresses in an upward and downward order (getting unidirectional and bidirectional bit bridging and address decode stuck-at and bridging faults); and will change addresses at speed (getting the address decode and row decode delay faults). Since this algorithm conducts a "read"-"address change"-"read," subtle delay faults associated with address changes and immediate writes are not adequately detected—an algorithmic pattern such as a galloping diagonal may be better for this detection.

4.13.2 Automatic Test Generation

To further reduce the cost of test, many testers provide a "Memory Test Function" that can generate algorithmic memory tests. This allows a sequencer to provide addressing and data information, instead of having to have tester patterns stored in pattern memory (for example, providing a 14N March C+ test on a single 4K (4096) memory requires 57344 clock cycles).

Since most algorithmic tests do not exercise all possible failure modes and fault classes, multiple algorithmic tests may be required. Many testers support most of the standard algorithmic tests such as: ping-pong, walking ones, galloping ones, galloping diagonal, checkerboard, address uniqueness, surround disturb, refresh, and alternate march tests with different data backgrounds (e.g., 0-F, 3-C, as well as 5-A).

4.13.3 BIST-Based Algorithmic Testing

Built-in Self-Test is also based on algorithmic test generation, however, the algorithm is embedded within the design. Since BIST requires additional logic to be added to the chip design to include the tester function, only one algorithm is generally designed and supported. Once an algorithm is embedded within the design, it can't be changed, so, to ensure that the most coverage is obtained by the designed-in algorithm, an analysis should be done based on the most common set of defects seen in the target fabrication plant, and based on the type of memory being tested.

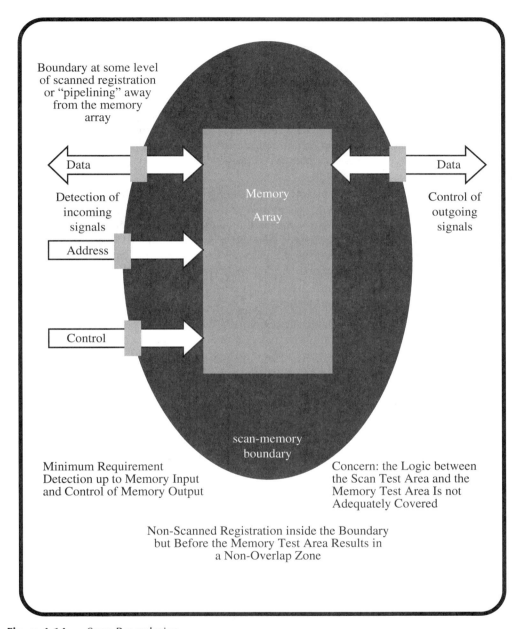

Figure 4-14   Scan Boundaries

4.14 Memory Interaction with Scan Testing

4.14.1 Scan Test Considerations

Most designs include some form of memory array or register file logic. When the majority of the general control logic is tested by scan, since it is not economically feasible to use scan to test the memory array (memory array cells are not scannable), then there is a concern that there may be a coverage loss between the applied memory test and the edge of the scan logic (see Figure 4-14). This happens because the data and control path near the memory may be one of the most critical timing paths in the chip, and to prevent adverse impact, the scan architecture may stop at some distance away from the memory (e.g., at least one level of registration—maybe more).

Different types of memory test architectures may result in different levels of coverage. For example, a direct memory access test strategy may fully cover the data and control path if the natural data and address paths are used. However, if a separate test bus is delivered to each memory and is multiplexed in right at the memory boundary, whether from a BIST controller or from a direct memory access controller, then the natural data and control pathways may not be observed and covered for fault detection.

4.14.2 Memory Interaction Methods

Several methods are used for the interaction of a memory with general chip scan logic. Each method has different trade-offs. The most common methods are "Black-Boxing," the "Fake Word," and "Memory Transparency." These will be explained, with each method's trade-offs, in more detail in sections 4.16, 4.17, and 4.18.

4.14.3 Input Observation

One of the main goals in the testing of an embedded memory area with scan is to provide fault detection as close to the memory as possible. If the memory does not have scannable registration within the cone of influence of the memory test architecture, then there is a gap or non-overlap zone between the two test arenas. To close the gap, the techniques applied are to scan-insert something in the memory test zone or to add observe-only registers to the target signals. This generally means connecting observe registers to the address, data, and control signals as close to the memory as possible.

4.14.4 Output Control

The memory can look like a large "blocked area" within a scan logic test area. Everything in the downstream shadow of the memory may experience a loss of controllability because the output from the memory array is unknown. Solutions are usually to: 1) fix the output of the memory to known logic values; 2) override the memory output with multiplexed known logic values; 3) make a transparent mode that applies the memory inputs to the memory outputs. These solutions can affect the memory output timing path. In some cases, more test control is needed to provide known outputs from memory-related logic such as the Tag or Cache-Hit logic (these signals can affect the coverage of combinational logic far from the memory local area).

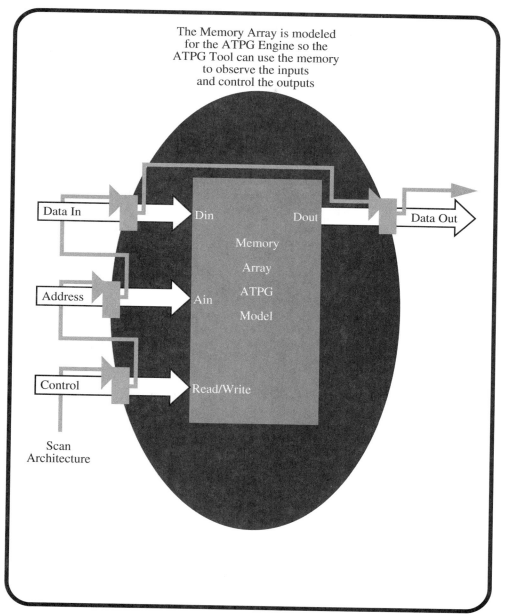

Figure 4-15   Memory Modeling

4.15 Scan Test Memory Modeling

4.15.1 Modeling the Memory for ATPG Purposes

When the purpose is to get as much coverage of the area between the scan architecture and the memory test architecture, allowing the ATPG tool to operate right up to, or even into, the memory is the most ideal situation (see Figure 4-15). From a design point of view, this results in minimal design effort, since the goal is to not do anything special with the memory for scan DFT because the ATPG tool can understand and operate the memory during scan testing. However, for this method, the requirement is to model the memory for ATPG purposes (and to apply an ATPG tool that can take advantage of it).

In some cases, requirements may dictate creating a memory operation architecture that the ATPG tool can deal with, or only supporting memory arrays that ATPG can operate. For example, the read_write control or the data output enable might be usable only by the ATPG tool if they are directly controlled by the tester (the signals are brought out to the package pins) or if the signals are directly controllable by scan registers. Also, a requirement might mandate that the clock that operates the memory be the same clock (or effective clock) that operates the scan domain feeding the memory data and control signals. However, some memory architectures are "immune" to ATPG.

4.15.2 Limitations

The modeling of the memory for the ATPG tool does not work if: 1) the ATPG tool does not handle memory interaction; 2) the memory is a type that is not handled by the ATPG tool; or 3) the changes to the memory architecture that are needed to allow an ATPG to operate the memory can't be supported (for example, they would affect the memory timing negatively). Limited options can be applied if any of these cases are true.

If the ATPG tool does not handle memory arrays and it is desirable to have the ATPG operate right up to, or into, the memory, then an effort needs to be made to allow the memory model to be replaced with flip-flops for the ATPG tool (i.e., the memory is represented as flip-flops in the netlist given to the ATPG tool).

If the memory is not easily modeled for ATPG or if the changes to allow the ATPG to operate the memory can't be supported, then a structure must be made around the memory that will meet the needs of the ATPG tool. For example, if the clock that operates the memory is a time-delayed version of the clock that operates the last set of registers feeding the memory, then the memory captures the same data as the registers. This outcome has the effect of making the last set of registers (and possibly scannable registers) transparent as far as the memory is concerned. A zero-delay ATPG tool has no way to model this, other than representing scannable registers as transparent latches. This may result in a loss of coverage somewhere else in the design. For test purposes, it may be necessary to synchronize the memory to a different clock during scan mode.

If a model cannot be used, then the memory may have to be removed from consideration by using one of the other "memory-handling" techniques.

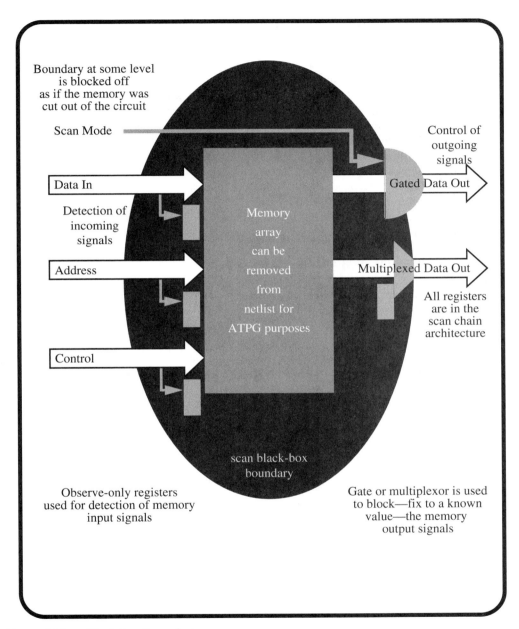

Figure 4-16 Black Box Boundaries

4.16 Scan Test Memory Black-Boxing

4.16.1 The Memory Black-Boxing Technique

The most common way to handle memory arrays embedded within scan logic is to "not handle them." In other words, the most common technique applied is to sample the input signals into scan capture registers and to force the output of the memory to some fixed value by direct gating. If control of the output signal to either logic value is needed, then an alternative to gating the output is to place a scan control register on the memory output through a mutlitplexor.

This method minimizes the scan delay impact on the input side, but may adversely impact the output delay on the memory output side by placing a gate or multiplexor in the path. Sometimes this delay can be minimized by conducting the gating within the memory output logic. If the output delay can't be tolerated, then a solution is to place the output control in a pipeline stage downstream from the memory array.

Generally, black-boxing provides a method for the ATPG tool to remove the memory from consideration by allowing it to be removed from the netlist (see Figure 4-16).

4.16.2 Limitations and Concerns

Problems occur with black-boxing the memory. The main problem is the extra logic placed around the memory to implement observability and controllability. Other problems occur if the memory array is a custom design and the custom design includes logic and scan chains. When this situation occurs, the netlist must be modified so that the memory is removed, but the logic and scan chains must remain. This modeling problem can be fairly difficult.

The black-boxing method also removes the natural timing involved with capturing input data into the memory array and with data launched from the output of the memory array. This may compromise any delay testing or timing verification associated with the memory during scan testing.

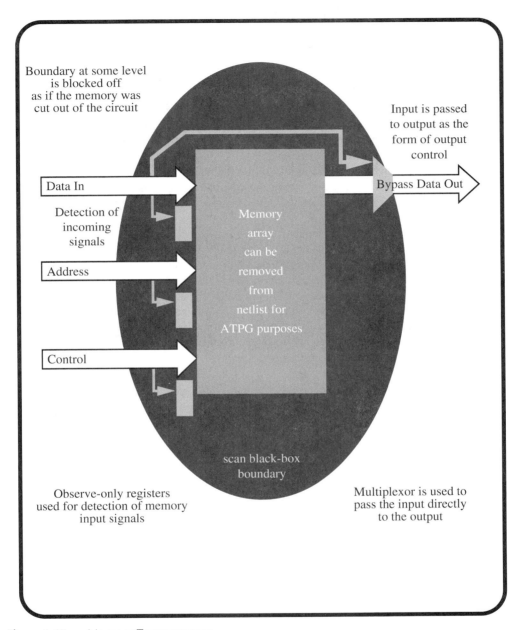

Figure 4-17 Memory Transparency

4.17 Scan Test Memory Transparency

4.17.1 The Memory Transparency Technique

Another common technique used on memory arrays embedded within scan logic is to bypass them (another form of black-boxing)—in other words, passing the input data directly to the output. This technique is generally applicable only for slow stuck-at testing, because when the memory is bypassed by simple connection, the natural timing is not maintained. If the scan chains are to be operated at speed, then the constructed bypass may pose a problem in that the loss of sequentiality (the memory) may make the data-input to data-output path a multicycle path—this can limit the frequency of scan testing.

For DC testing, however, this technique allows the ATPG engine to control the data downstream from the memory by applying data on the upstream side of the memory. The upstream logic is observed by the scan registers on the downstream side as well (see Figure 4-17).

4.17.2 Limitations and Concerns

The problem with direct transparency is that the bypass removes the natural sequential effect of the memory. The bypass path may create a multicycle path. For a zero delay ATPG tool, this situation will create a scan test that cannot operate at the rated frequency of the device.

Also, the memory input and memory output paths are usually some of the most critical paths in the design. If a BIST memory test or a memory test access from a non-functional pathway is provided, then the functional critical paths may not be tested as part of the memory testing. In this case, the scan test may be the only way to measure these critical paths. If so, then scan testing can't be allowed to bypass the memory arrays.

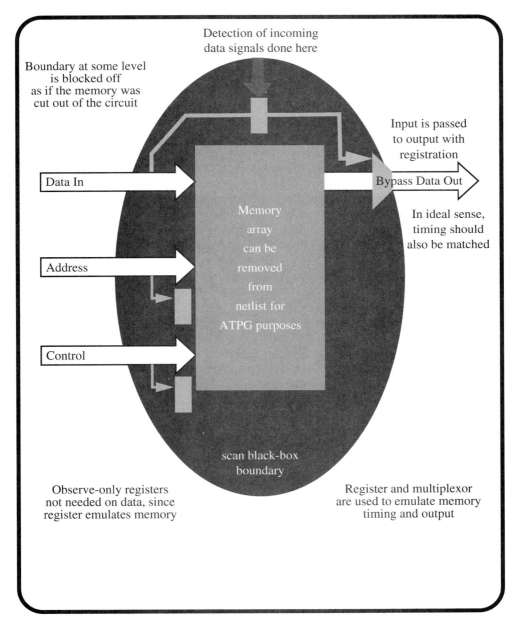

Figure 4-18 The Fake Word Technique

4.18 Scan Test Memory Model of The Fake Word

4.18.1 The Fake Word Technique

The transparency technique is one way to allow the applied input to the memory to provide the necessary output data from the memory. However, passing the input to the output does not maintain natural timing (the sequentiality of the memory is missing). This lack limits the scan testing near the memory to some frequency that is less than at-speed, and may limit all scan testing. If the chip-testing goals include scan testing for AC coverage (transition and path delay), then the transparency method may not be sufficient. To more emulate the memory pathway, sometimes a register is placed in the transparent data path (see Figure 4-18). However, if the AC goals also include testing the pathways into and out of the memory, and with proper timing, then a real memory word should be emulated.

Emulating the memory array's role in the bypass path can be implemented in several ways. The memory can be bypassed with a register or by reserving one actual memory location (address) for test. Whenever the part is in scan test mode, the bypass logic should be enabled—this may mean selecting the register path or forcing the memory address to always select the "fake test word."

This method, when coupled with an ATPG model of the memory, allows the scan architecture to be used to test the speed paths into and out of the memory, and no non-overlap zone exists when a memory test architecture is applied.

4.18.2 Limitations and Concerns

This technique is one of the best methods, and when coupled with an ATPG tool that allows a memory model, then ATPG can be accomplished and the memory array can still be removed from the ATPG netlist. However, this method requires some design interaction, and depending on how accurate the timing of the "fake" circuit needs to be, some physical design work and analysis on the timing habits of the circuit may need to be done to support a "realistic" timing measurement.

4 • Memory Test Architectures and Techniques

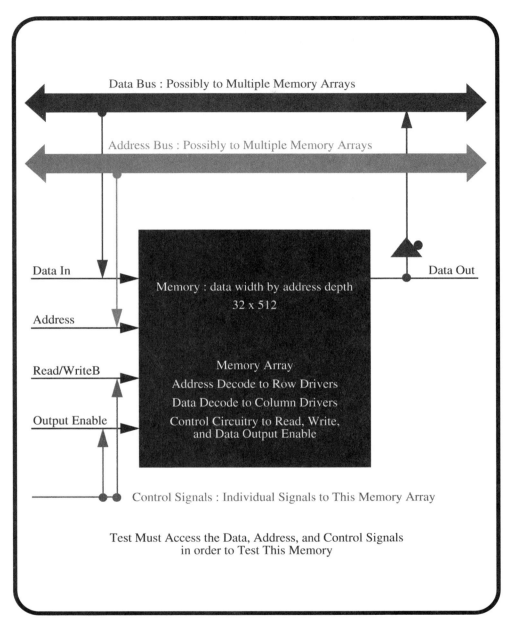

Figure 4-19 Memory Test Needs

4.19 Memory Test Requirements for MBIST

4.19.1 Memory Test Organization

When memory arrays are embedded within a chip, the memory data, address, and control signals must be accessed for memory testing (see Figure 4-19). The form of testing applied is "algorithmic" in nature in that sequences of reads and writes with various data background pairs (e.g., 0-F, 5-A, 3-C, 9-6) are applied in order to exercise and detect certain fault and defect classes.

The testing may be applied by using a direct connection to the tester (the algorithmic test is generated within the tester), or it may come from a source within the chip (the algorithmic test is generated within the chip). If the ability to conduct testing is embedded within the chip, then the test architecture is generally referred to as a Built-in Self-Test, or BIST. When the BIST is applied to a memory, then it is called memory built-in self-test or MBIST. If several separate memory arrays are supported within the chip, then multiple, separate, embedded test structures may have to be supported since different memory arrays may have different data width and address depth requirements.

4.19.1.1 Test Address Decoding

To effectively test the memory, the test architecture must have the ability to sequence through the complete address space of each memory. The sequencing itself must allow the application of standard test algorithms. This requires that the memory addresses be controllable by the test logic so that they may be incremented or decremented as needed.

4.19.1.2 Test Data

The data applied to each memory must be organized standard data patterns to support defect coverage. These patterns are generally complementary pairs such as 0-F, 3-C, 5-A, or 9-6; however, some algorithms apply data such as a filling 1 (a 1 that increasingly fills the data word with each write) or a walking 1 (a 1 that seems to walk across a data word). The input data bus must be controllable to allow the application of these test data and the output data bus must be observable to verify the results.

4.19.1.3 Control Signals

Any signal that controls the ability to read, write, program, erase, or present data to the output needs to be controlled by the test sequencer.

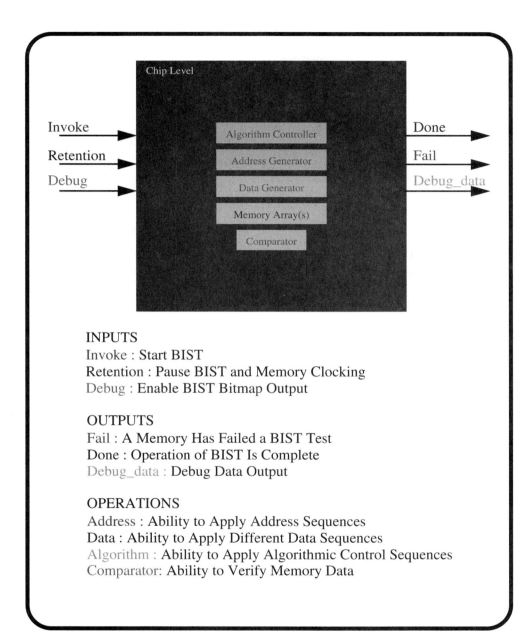

Figure 4-20 Memory BIST Requirements

4.20 Memory Built-In Self-Test Requirements

4.20.1 Overview of Memory BIST Requirements

In order to move the tester into the silicon, several test requirements must be fulfilled. At the chip level, some form of test logic will be embedded, and this test logic must meet the test requirements applied to all chip logic. The minimum requirements for an embedded memory BIST are to be able to apply at least one algorithmic test to the full memory space with one complementary data background pair. A method must also exist to assert the MBIST and to report MBIST status at the chip level. Some IC organizations have a higher test standard and require the ability to conduct a memory retention test and to output some form of debug or bitmap data (see Figure 4-20).

4.20.1.1 Interface Requirements

There are many options for providing the control and reporting structure for internal embedded Memory BIST. Integration decisions and trade-offs will establish the exact architecture—whether it is dedicated chip pins, borrowed chip pins, an IEEE 1149.1 instruction, or some other form of test controller. At the chip level, signals need to exist for one or more invoke assert, retention assert, and debug assert signals (unless the normal operation of the BIST is with automatic retention and debug enabled; then only an invoke signal needs to be supported).

The same integration decisions and trade-offs will establish the reporting architecture. At the chip level, the BIST logic needs to supply an indication of being done and an indication that something has failed. Providing debug data may also be a requirement.

4.20.1.2 Operations

The embedded Memory BIST logic needs to support, for any given memory, the ability to sequence the memory with a known memory test algorithm. This requires that an algorithm sequencer, or state machine, be provided and that this algorithm sequencer have control over an addresser and a data generator. This setup would allow the embedded logic to implement, for example, a March C type test that would Read(data), Write(complement), and Read(complement) up and down through the memory space. To ensure that the read and write operations are correct, a comparator must be used.

Extra features for comprehensive testing would be to allow the memory and the BIST to pause so that a retention test can be done, and to provide output debug data to assist in finding the cause of a failure.

4.20.2 At-Speed Operation

One of the main goals of memory testing is to ensure the access and recovery time of the memory arrays. This can be done only if the MBIST allows at-speed operations to occur (operation at the rated clock frequency). Because of this requirement, most BIST logic is developed as HDL and is brought through synthesis with timing constraints to ensure that it operates at speed. BIST can be gate-level netlist inserted but only if coupled with some form of timing analysis, since memory pathways are generally very timing-critical.

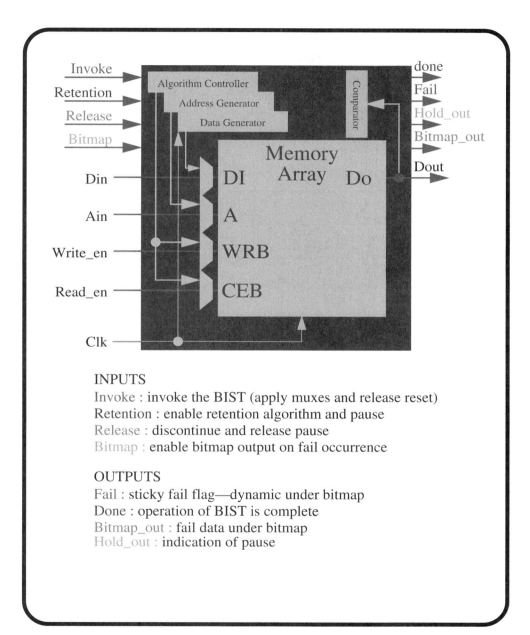

Figure 4-21 An Example Memory BIST

4.21 An Example Memory BIST

4.21.1 A Memory Built-In Self-Test

An example Memory BIST appears in Figure 4-21. This example shows a memory with associated BIST circuitry that meets the test goals. At the memory level, this MBIST includes a signal interface for control and data and for output reporting. Within the memory BIST unit are the functional elements that make up the BIST operations.

4.21.1.1 Memory BIST Interface

As can be seen in Figure 4-21, this memory unit requires four extra input signals and four extra output signals to enable communication with the BIST circuitry. In this case, the invocation of BIST is done by first applying multiplexors to take over the data, address, and control memory signals when "Invoke" is asserted, and then by releasing the reset on the next clock cycle. The "Retention" input signal allows data retention testing (a pause of the BIST and Memory operation), and the "Bitmap" input signal enables the presentation of bitmap data on the debug output signal, "Bitmap_out."

On the output side, the "Fail" signal asserts for any failure (data miscompare), and the "Fail" signal should be a "sticky bit" to let a tester poll the output signal at its convenience instead of having to strobe and compare every cycle (this allows the BIST pattern in the tester to be only a clock loop and not a pattern in test memory). The same requirements are true for the "Done" signal.

4.21.1.2 Algorithm Sequencers

The main component of BIST operation is the algorithm sequencer. This is the hardwired pre-selected algorithm, or algorithms, that can be applied to exercise and detect targeted defect and fault classes. Some of the most popular memory BIST algorithms applied are in the March family (March, March II, March C-, March C, and March C+). This is a general-purpose set of algorithms that have high defect coverage and a fairly low test time impact (7N to 14N—where N is an operation per word). The algorithm is a series of reads and writes coupled with address selection and data application (refer to Section 4.13).

4.21.1.3 Address Counters

A simple address counter is needed that has the ability to count up the address space and down the address space. The direction and addresses applied at any given time are controlled by the algorithm sequencer. Linear Feedback Shift Registers (LFSRs) can also be used as address generators, and they usually have a lesser logic and area impact as compared to a standard counter.(Note: An LFSR and the exact inverse LFSR may be needed to transverse the address space in forward and exact reverse order.)

4.21.1.4 Data Generators

The applied data for memory testing are usually complementary pairs (at least for the March algorithms). These data values can be generated by multiplexors with hard-wired logic values on the inputs, or the data values can come from a counter or register bank.

4.21.1.5 Comparators

The verification that a BIST memory test has passed is to verify that a read operation returns the correct value. This is most efficiently accomplished by comparing the memory output to some stored value. Since the BIST controller creates the data to be read into the memory array, then using this value to compare to the read data coming out of the memory is a simple operation. Other less efficient methods include signature collection and bringing the read information out to the edge of the chip.

4.21.2 Optional Operations

In this case, retention is applied by asserting the "Retention" signal at any time during the BIST operation. The retention operation may be directly controlled by the assertion of the "Retention" signal, and then the pause operation is ended by asserting the "Release" signal, or the memory may include a "self-pause" operation that is enabled while the "Retention" signal is asserted, and the self-pause can be ended by asserting the "Release" signal.

Bitmap data is very important in identifying the reason for a fail—during production test, identifying that the memory has failed is important—however, for yield enhancement or memory characterization, having access to the address and data that is being applied to the memory is a critical necessity. In this example BIST, the Debug data may be provided on the "Bitmap_out" signal with each data miscompare (fail), or the "Bitmap" signal may enable the data output, both passing and failing.

4.21.3 An Example Memory Built-In Self-Test

For deep submicron design, the number of wires routed to and from the BIST controller is a major concern to the integrator (especially with many distributed memory arrays). Another concern is having to simulate the memory BIST to find the cycle where a tester pause must be applied. A way to solve this problem is to support a comprehensive BIST controller that supports a combined test and retention algorithm.

The combined algorithm allows the memory to pause when the data included within the memory is defined. This eliminates having to route individual "assert hold" signals to each memory. The global "Retention" signal is designed to tell the BIST controller to conduct a self-pause when the memory has been initialized. The input signal known as "Release" is used to allow the BIST to continue. An optional "Hold_out" output can be used to coordinate many BISTs that are paused, or can be used to indicate to the tester when the global "Release" should be asserted.

The self-pause algorithm is one that uses the natural stopping point of the last address to indicate to the BIST controller that it should place itself into pause mode. The ideal data background is to map the physical memory locations so that the data effectively places an A in the even address locations and a 5 in the odd address locations to produce a checkerboard pattern in the physical memory (and odd=A, even=5 for the complement).

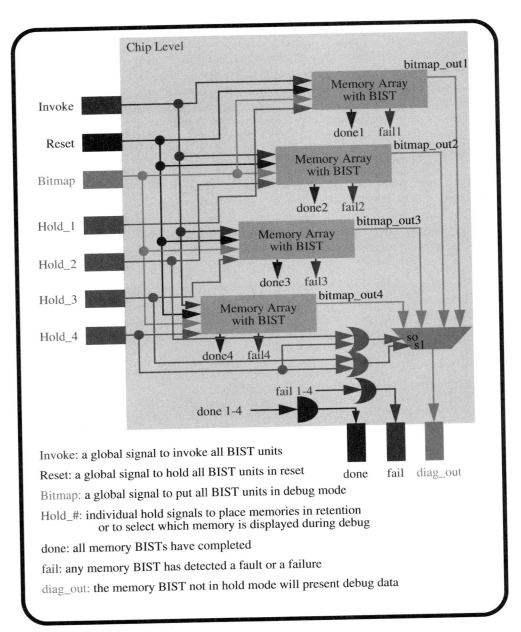

Figure 4-22   MBIST Integration Issues

4.22 MBIST Chip Integration Issues

4.22.1 Integrating Memory BIST

Many trade-offs need to be considered when multiple embedded memory arrays are to be placed within a single chip. The major decision is whether to support one BIST controller for all memory arrays or to support one BIST controller for each memory array (or some mixture of the two). This decision is usually based on the process geometry, the number of memory arrays to be integrated, and the targeted placement of the memory arrays (see Figure 4-22). If the process geometry results in feature dimensions below .5 micron, then routing and route delay is the big concern. If many memory arrays are to be integrated and they are dispersed widely across the chip, then routing is again a concern. A design where routing is a concern would benefit from a BIST controller per memory, or what is termed "a BISTed memory."

As another example, if few memory arrays are supported or if all the memory arrays are placed in one location (e.g., all on one side of the chip), then supporting a single BIST controller with a single routing interface may be a more optimal solution.

4.22.1.1 Single Memory BISTs

When the process geometry is above .5 micron, then gate count or cell area is a dominating concern. Generally in this case, the number of memory arrays is limited. In this situation, it may be better to support a single BIST controller and to route busses to each memory. If the process geometry is below .5 micron and route delay is the dominating concern, then a single BIST controller is not recommended unless all the memory arrays are placed close together physically; for this case, a single BIST controller can be supported, because the busses do not have to be routed very far or because a common bus structure to all of the memory arrays can be created.

A single BIST controller means that only one set of logics is required to implement the address generator, data generator, and algorithm sequencer. In most designs, each memory array must be tested individually and "one after the other," usually depending on the bus structure (but one-after-the-other may be a test time concern).

4.22.1.2 Multiple Memory BISTs

When one controller per memory array is supported, then BIST integration is just the tying together of all the invoke signals, the ANDing of the "dones," and the ORing of the "fails." Some cleverness may need to be applied to the "holds" and the "bitmaps" to minimize the impact of routing within chip goals. For example, 25 individual memory arrays with individual BISTs could require 25 hold signals and 25 bitmap output signals. The routing of this many signals may be similar in cost to providing a "direct memory access" test architecture (e.g., a 32-bit chip pin interface).

One other consideration with many multiple memory BISTs is the power rating. If all the memory arrays are active during BIST testing and they all potentially could commit a read operation on the same clock cycle, then a power consumption spike that exceeds the package rating or the silicon power grid capabilities may occur.

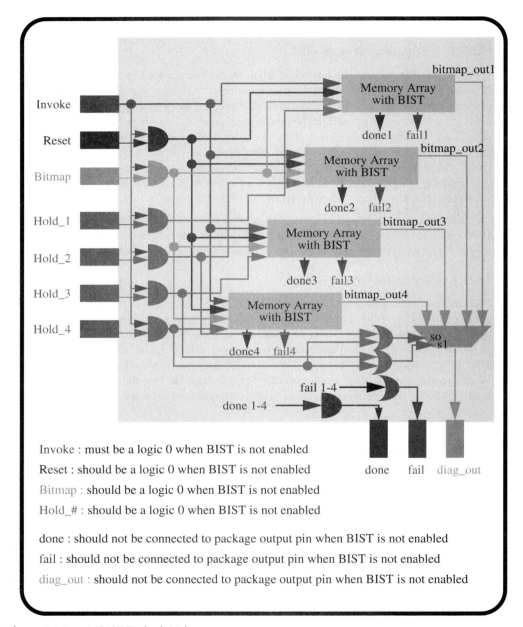

Figure 4-23 MBIST Default Values

4.23 MBIST Integration Concerns

4.23.1 MBIST Default Operation

When the BIST logic is not being used, an effort needs to be made to ensure that it is quiescent and does not interfere with functional operation. The term "interfere" actually means that the BIST logic, which is considered test-only logic, should not affect the functional operation, the chip's power consumption, or the timing and loading on critical signals or interfaces. When integrating a single MBIST, or several MBISTs, into a chip design, ideally, the application of default values should result in minimizing power consumption and should ensure that there is minimal impact to all borrowed interface pins or critical timing paths (see Figure 4-23).

4.23.1.1 Default Control Signals

When Memory BISTs are integrated into an overall chip design, an effort needs to be made to ensure that the test logic remains quiescent when the chip is not in a test mode. It is not enough for the BIST logic to not operate on the memory array (e.g., the multiplexors that take over the memory apply the functional path when not in a test mode). In the ideal sense, the BIST logic will not operate, will consume minimum power, and will present minimum loading to critical signals when it is not used.

The most common technique used is to define default control signal values and to factor a test mode signal into gates that distribute these signals to the various MBISTs. For example, when the chip is in functional mode or in a test mode that does not use the memory such as scan test mode, then the gated control signals will de-assert the multiplexors, will place the BIST into reset, will de-assert the debug function, and will assert the hold function (in case a fault in the reset logic allows the BIST to operate). By using a gated system, the chip user does not have to dedicate input pins for test and then require that certain constrained logic values be applied to those chip package pins when they are integrated into a board or system design.

4.23.1.2 Default Output Interface

Similar rules apply to the MBIST test-only output signals. When the chip is not in a memory test mode, then the output signals should not consume power and should not have a negative impact on chip functional output pins, especially if the BIST control interface and the BIST data output interface comprise borrowed package pins. The output signals exiting the MBIST should drive a known fixed value when the chip is not actively using the MBIST.

4.23.1.3 Quiescence

In extreme cases, the MBIST logic must be rendered absolutely quiescent. This condition requires gating off the clock to the BIST logic. This may be difficult if a generic place and route tool is used and the chip flip-flops are placed on the branches of a clock tree. In this case, the MBIST flip-flops should be made with local gating, and the test mode signal should be routed to each one.

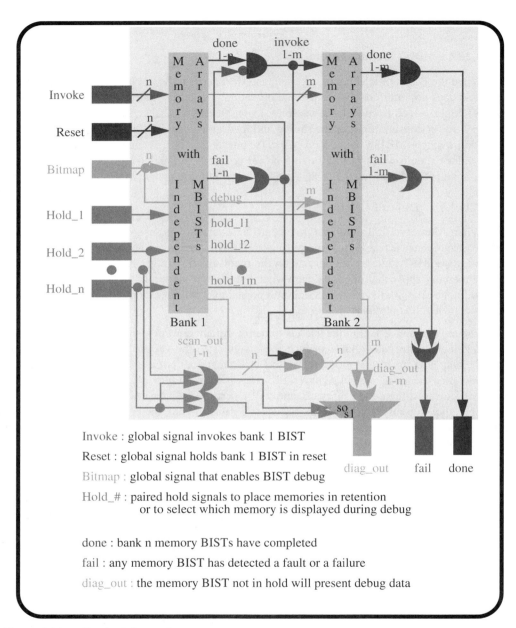

Figure 4-24   Banked Operation

4.24 MBIST Power Concerns

4.24.1 Banked Operation

One of the advantages of MBIST is that all memory arrays can be tested simultaneously, which reduces test time. However, if the chip has integrated a great number of memory BISTs, then one possible disadvantage that may occur is that the power rating of the package (or silicon) may be exceeded if all memory arrays are tested in parallel.

The solution to this problem is to limit the number of memory arrays that are tested at any given time. This technique is known as banking (see Figure 4-24). A bank is several memory arrays with BISTs that are designed to be tested in parallel. Each bank can be asserted individually to allow testing to be conducted on just those memory arrays. In the ideal sense, the best technique to apply is to allow each memory bank to control when the next memory bank starts testing. This is done by ANDing all the DONE signals and using this signal as the next INVOKE signal. The "self-invoking banking" technique is safer than having multiple invoke signals—multiple invoke signals may allow an end user to accidentally assert more than one bank and possibly exceed the power rating of the chip package (or exceed the on-silicon power delivery capability, resulting in "brown out").

An analysis must be done during the chip development stage to assess which memory arrays and how many should be included in a bank. This is first done by establishing a power specification such as 50% to 75% of the Max package rating. The rating needs to include some guardband, since some logic may be active during the BIST operation other than just the memory and BIST logic. Next, each memory array needs to have a worst-case power value calculated for it. In general, the worst-case operation is a read, and the power consumption depends on the aspect ratio and the access time of the individual memory (these are physical aspects that need to be understood when the MBIST architecture is being developed).

The power rating is usually given in terms of milliwatts per MegaHertz. The MegaHertz used for the calculation should be the Max operating frequency of the chip. During the design phase, the assumption that test will be applied at a lesser frequency may not always be a safe assumption, since testing may be applied by a different organization (and the test plan may not always be communicated effectively). Also, at some future point in time, the MBIST may be used "in-system" as a real-time confidence test—in this case, the chip may be operated at-speed during the test.

In any case, the power is calculated as if the there is a possibility that all the memory arrays in a bank could have a worst-case operation in any given clock cycle. So a per-cycle worst-case operation at the Max frequency should be calculated for each memory. Then the memory arrays should be grouped by the "Test N" rating with a thought for the least overall chip test time. For example, if six memory arrays have similar power consumption, only two memory arrays can be tested at any one time, and three of the memory arrays had short test times, and the other three had long test times, *then* it would be foolish to include the longest to test memory arrays in each bank. The overall test time would be shorter with the long tests grouped together.

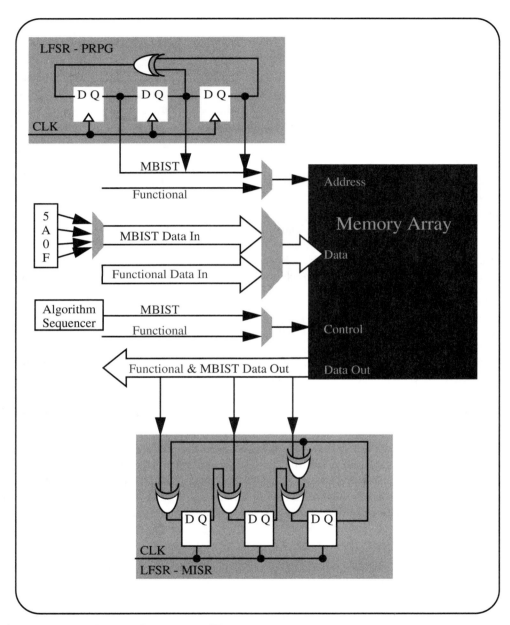

Figure 4-25 LFSR-Based Memory BIST

4 • Memory Test Architectures and Techniques

4.25 MBIST Design—Using LFSRs

4.25.1 Pseudo-Random Pattern Generation for Memory Testing

Relying on an "in-order counter" for MBIST is not always the best way to conduct an embedded memory test. Alternate types of BIST controllers attempt to minimize the logic overhead of providing BIST. For example, one method is to rely on the Pseudo-Random Pattern Generator Linear Feedback Shift Register (PRPG-LFSR). The PRPG-LFSR is a pseudo-random pattern generator (for which the count order is seemingly random, depending on the initial value or seed, but for a given seed it is a repeatable sequence). If the LFSR is based on a primitive polynomial, it can be viewed as a counter for the whole address space (except address 0), but not in the traditional count order. The LFSR generally has a smaller gate impact than a traditional in-order counter (the same number of flip-flops, but the next-state logic is just a few exclusive-ORs).

The PRPG-LFSR can be used as an address counter for MBIST algorithms on one condition—if the PRPG-LFSR is used to conduct the forward count of an algorithm (such as a March type test), then the "inverse LFSR" must be used to create the backward count (see Figure 4-25). This guarantees that the reverse order is applied exactly reverse to the forward order, that all locations are tested, and that the bridging algorithms are satisfied.

Since the data for any memory test algorithm is either a constant logic 1 being replaced with a constant logic 0, or a complementary pair (such as 5-A), then there is no real advantage to using an LFSR to create the data. Similarly for the read-write control, there is no advantage to applying control in a random or pseudo-random sequence.

4.25.2 Signature Analysis and Memory Testing

There are several methods used to verify that the memory output is correct. Each method has its trade-offs. One real-time method is to use a data comparator. This is a good minimum hardware method if there is not a problem with observing the failure when it occurs. For some testers, it is more economical to check a pin or value after the test is completed (for example, as a result of the "done" signal transitioning), instead of checking for the "fail" signal during each clock cycle.

One method uses an LFSR to collect and compress each read. The LFSR is used as a data compressor (concurrent divider) that is enabled to capture data after each requested read. The process of capturing each read and compressing it based on some polynomial is known as *signature analysis*.

4.25.3 Signature Analysis and Diagnostics

A real-time fail signal allows for some diagnostic capability (when the fail occurs, the fail signal transitioning will identify the failing word, and techniques can be applied to present the data bit within the word as well). The real-time failure indication can be used to provide a bit-map that will allow for "multiple clue analysis."

Diagnosis of a failure or failures from a compressed LFSR-generated signature is possible, but it is only straightforward under the single fault assumption. For a single failure, a signature dictionary must be created where every possible single-bit failure is calculated and stored (this creation is generally done by simulating every possible data sequence that feeds the LFSR and is done with one installed error for each simulation). Real tester failures can be compared to the dictionary to identify the error.

The signature dictionary could become an intractable problem in the presence of more than one failure. For example, just considering a simple 8 word by 8 bit memory, and the single fault assumption, there are 64 possible single bit failures (there are 64 bits in an 8x8 memory array). This means that there are 1 good signature and 64 possible error signatures, so only 65 signatures need be calculated and stored.

However, with just 2 bit failures considered, for each of the 64 single bit failures, there are 63 possible next bit failures, or 64 times 63 possible error signatures (4032 signatures). To calculate (simulate) a signature for every possible combination and number of errors that the 8-by-8 memory could support, then the number of simulations that need to be run to fill the signature dictionary is 64! (64 factorial). However, a problem could crop up with aliasing. Combinations of errors may create a self-repairing signature. This can also evidence itself as repeated error signatures. As an example, if the LFSR for the 8x8 memory array is just an 8-bit LFSR (one bit per data line), then there are $2^N = 2^8 = 256$ possible values in the LFSR—given just the 2 bit errors, there are 4032 possible signatures—so there will be many repeated signatures. It would take a 12-bit LFSR with 4096 values to reduce the possibility of aliasing to a probability that no repeating signatures will occur for the 2 bit error case. Imagine the size of the LFSR to handle 64!—a number greater than 1.2×10^{89} (hint: it is around 295 bits, which makes the LFSR larger than the memory).

For larger memory arrays (for example, 4 Mb) the calculation can become totally intractable (e.g., 4 million factorial)—and even if the simulations could be done, the size of the LFSR needed to reduce aliasing would not be cost effective.

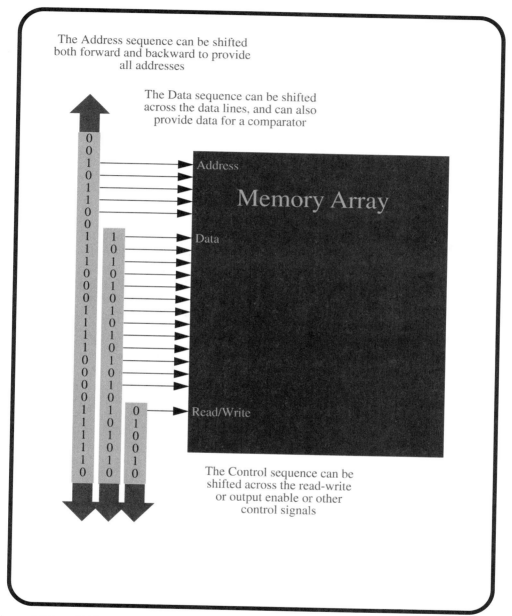

Figure 4-26 Shift-Based Memory BIST

4.26 Shift-Based Memory BIST

4.26.1 Shift-Based Memory Testing

If the memory has registered data, address, and control lines and if placing a multiplexor in the path of these signals is not the preferred BIST solution, then the natural registers can be converted to scan shift registers. These shift registers can be used in such a way that they can implement the memory test algorithms (see Figure 4-26). For example, if a particular sequence of 1s and 0s is passed across the address lines, then every word can be addressed (e.g., the sequence represents a minimized count order sequence). This sequence must be passed across the address lines in reverse to implement the reverse order count.

While the address sequence is being passed across the address lines (with pauses to emulate the applied algorithm), then the data applied can be a serial sequence of complementary data. For example, if the shift data is a stream of 1, then 0, then 1, etc., then the applied algorithm, while addressing forward through the memory, may be read a 5, write an A, read an A, and when the address is reversing through the memory algorithm, the data would be read an A, write a 5, read a 5.

The sequence control necessary for these test operations is generated in two areas: first, the read-write is done by a serial bit stream such as Read=0 and Write=1, so 010010 would represent a Read-Write-Read, Increment Address, Read-Write-Read; and second, the address data stream must hold the value of any address for at least three clock cycles so that a Read-Write-Read sequence can be applied continuously to any given address. This sequence control requires that the BIST controller also control the relative shift rates (clocking) to the various different shift registers.

4.26.2 Output Assessment

The scan-based memory test scheme can be used to apply the algorithmic test, but the output assessment that should be used needs to be the same one used for more traditional BIST. The capturing of the output as a scan word and shifting out the data is not the most effective method—it is much more effective to use a comparator or a parallel signature analyzer.

4 • Memory Test Architectures and Techniques

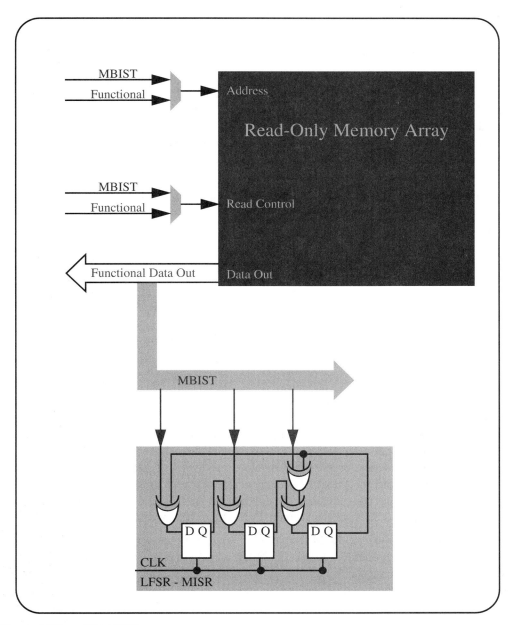

Figure 4-27 ROM BIST

4.27 ROM BIST

4.27.1 Purpose and Function of ROM BIST

A memory BIST for a Read-Only Memory has two purposes: one is to verify the data that is stored within the ROM, and the other is to ensure that no defects exist that would alter or destroy that data when a read operation is conducted (the destructive read). The reasons for embedding a test function within the chip are the same as those described for RAM BIST: to provide a reduced test interface to an embedded memory function; to embed a high-coverage test function that will provide consistent results regardless of the target tester; and to allow test time reduction by providing embedded tests that can be operated simultaneously.

The ROM BIST is different from the RAM BIST in that the ROM is pre-programmed and can only be read (see Figure 4-27). The advantage of the read-only memory is that the BIST algorithm must only read the data. The difficulty is that the read data must be compared to known data—and the data is diverse data, not the known background data pairs used in RAM testing. It is not efficient to place an extra copy of the ROM data into the BIST controller to use as a "golden standard" for comparison, so a different method must be used. The usual method for ROM verification is to use a compression scheme and to then compare the compressed data, or signature, to a golden compare value created at the time the ROM data is programmed. The compression scheme is based on the well known Cyclic-Redundancy Code (CRC) methodology and is based on the use of the Linear Feedback Shift Register (LFSR). An LFSR that has as its inputs the output data bus from a memory is generally referred to as a Multiple Input Signature-Analysis Register (MISR).

The MISR-LFSR is a shift register that has the data connection from the last (rightmost) bit brought back to various bits along the length of the shift register through exclusive-OR gates (the memory databus is also brought into the LFSR through exclusive-OR gates). The bits that receive the feedback are chosen from well known mathematical tables of "prime" or "primitive" polynomials. The initial state of the LFSR before it captures any data is a state known as the "seed." The operation that the MISR-LFSR conducts is one very similar to division by a prime number—where the input data stream is the dividend, the polynomial is the prime divisor, and the state of the MISR-LFSR after each capture cycle is the remainder.

The algorithm applied to ROM testing is different from the algorithms applied to RAM testing. The first pass through the ROM should be to begin at address 0, to read each address location as the test increments through the address space, and to capture this data in the MISR. This first pass verifies the program data and the address decode.

The second pass through the memory should start at address max and decrement through the address space to verify that the first-pass read operations did not corrupt any data with a destructive read operation. A third pass beginning at address 0 and incrementing through the address space again should be conducted to verify that the reverse-order read did not exercise an address order or sequence direction-dependent destructive read operation.

4 • Memory Test Architectures and Techniques

Note also that a retention test is not required for a ROM, but if the ROM is to be tested in the same environment with other memory arrays (RAMs and DRAMs), then including the ability to be placed in "hold/pause" mode will allow the ROM test to operate simultaneously with other MBISTs and to gracefully remain quiescent during retention testing or during an Iddq measurement.

4.27.2 The ROM BIST Algorithm

For completeness, an example ROM BIST algorithm is included. There should only be one basic algorithm within the ROM MBIST controller, a structural defect-based ROM March test algorithm. However, this algorithm may be modified slightly under bitmap and retention testing conditions.

When the "Invoke" signal is asserted, the ROM March sequence should begin, which conducts at least three operations (reads) per memory address location (logical word). The ROM BIST should be designed such that the golden comparison signature, or the signature modifier that sets the MISR to a known value such as all 0s, is stored in the last address (address max), so this value should not be read during the testing except as the very last operation (this requirement makes the calculation of this value easier, since it will not factor into the test).

The algorithm sequence is:

- Begin at first address location (address 0)
- Read (Data) - Compress (Data)
- Increment Address and repeat for entire address space (up to address max-1 if the signature compare value is stored in address max)
- Begin at last address location (address max-1 if the signature compare value is stored in address max)
- Read (Data) - Compress (Data)
- Decrement Address and repeat for entire address space up to address 0
- Begin at first address location (address 0)
- Read (Data) - Compress (Data)
- Increment Address and repeat for entire address space (including address max)
- Compare Signature to Stored Signature Value
- Assert "Done" and Assert "Fail" if necessary

4.27.3 ROM MISR Selection

The variable aspect of ROM BIST is the databus size (word width), and hence the supported prime polynomial for the MISR can also be viewed as a variable. The other variable is the method of comparison of the signature to a stored value. The preferred MISR selection method is to support a MISR that is at least the width of the databus where the MISR (and its polynomial) should be selected from a lookup table of prime polynomials. The prevention of aliasing, or an attempt to increase the uniqueness associated with the signature dictionary, may require using a larger MISR, since the probability of aliasing decreases as $1/2^n$ (where n is the number of bits in the MISR).

4.27.4 Signature Compare Method

The preferred signature comparison method is to provide a "last read" value and to program this value in the address max location. The applied BIST algorithm should read the entire memory three times, except for address max, and address max should be read only as the very last read operation. The value placed in address max should set the MISR to a known signature value such as all 0s or all 1s so that the comparator can be an N-input NAND-gate or an N-input OR-gate function. This type of comparison results in the exact compare being a logical 0, and any mismatch results in a logical 1. This signal can directly represent the "fail" flag.

Memory Testing Fundamentals Summary

Memory Testing Is Defect-Based

Memory Testing Is Algorithmic

Different Types of Memories—Different Algorithms

A Memory Fault Model Is Wrong Data on Read

Memory Testing Relies on Multiple-Clue Analysis

A Memory Test Architecture May CoExist with Scan

A Memory Can Block Scan Test Goals

Modern Embedded Memory Test Is BIST-Based

BIST Is the Moving of the Tester into the Chip

BIST-Based Testing Allows Parallelism

Parallel Testing Impacts Retention Testing

Parallel Testing Impacts Power Requirements

Parallel Testing Requires Chip-Level Integration

Figure 4-28 Memory Test Summary

4.28 Memory Test Summary

The embedded memory is becoming common on modern ICs, so common, in fact, that the modern integration technique is to support tens of distributed memory arrays on a single die. These memory arrays fall into several type groupings such as ROM, SRAM, DRAM, EEPROM, and Flash. There are several design issues and trade-offs involved with creating a test architecture for these embedded memory arrays, and in conjunction with other chip test logic such as scan. Figure 4-28 shows the memory test information, test methodologies, and test techniques that should have been learned in this chapter. In greater detail, they are:

- Memory Arrays are regular structures and are tested against defect models directly (instead of fault models, as logic is).
- Being a regular structure allows memory arrays to be tested algorithmically.
- The memory fault model is "any read operation resulted in the wrong data."
- There are many different types of memory arrays, and different types of memory arrays should be tested with different algorithms to target their individual defect types.
- Memory defect, fault, and failure diagnosis is based on multiple-clue analysis that relies on "bitmapping" several failures detected by different applied sequences.
- Memory test architectures may co-exist with other test architectures such as scan, and if not analyzed carefully, the memory test architecture may block scan test goals (and vice versa).
- Methods can be applied to allow the memory and scan test to support test goals in both environments—techniques such as black-boxing, memory bypass, and the fake word allow scan to verify the memory pathways, and allow the memory to assist in scan testing.
- Built-In Self-Test is becoming the solution to multiple embedded memory arrays, and this allows test time reduction by allowing many memory arrays to be tested simultaneously.
- BIST allows test parallelism, which reduces test time, but test parallelism may complicate the ability to conduct economical retention testing and memory characterization/debug operations.
- Test parallelism may also adversely impact chip power requirements—an additional analysis and chip-level MBIST architecture may be required that separate the parallel operation of many memory arrays into banks.

4.29 Recommended Reading

To learn more about the topics in this chapter, the following books are recommended:

Abromovici, Miron, Melvin A. Breuer, Arthur D. Friedman. *Digital Systems Testing and Testable Design*. New York: Computer Science Press, 1990.

Dillinger, Thomas E. *VLSI Engineering*. Englewood Cliffs, NJ: Prentice Hall, 1988.

van de Goor, A. J. *Testing Semiconductor Memories: Theory and Practice*. Chichester, U.K.: John Wiley and Sons, 1990.

CHAPTER 5

Embedded Core Test Fundamentals

About This Chapter

This chapter contains the Embedded Core Test Fundamentals and Core Design-for-Test Techniques designed to teach the basic and practical fundamentals of embedded core test and design-for-test. This chapter is a practical treatment of the concerns, issues, and design-for-test techniques involved with the development and subsequent integration of a reusable core.

This chapter includes material on: the time-to-market, design-for-test, and cost-of-test issues involved with developing a reuse core; embedded core test requirements; the reuse core test interface; embedded core test wrappers; embedded core scan-based and BIST-based testing; chip-level DFT development with consideration to a core or multiple cores; individual core testing in isolation; chip-level testing aside from embedded cores; chip-level testing of chips with embedded cores and embedded memory arrays; embedded core requirements; reuse vectors; and the chip-level test programs that include reuse vectors.

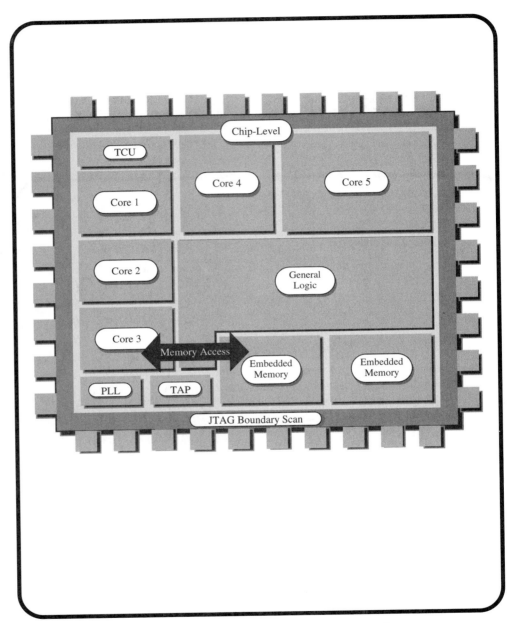

Figure 5-1 Introduction to Embedded Core Test and Test Integration

5 • Embedded Core Test Fundamentals

5.1 Introduction to Embedded Core Testing

5.1.1 Purpose

The Embedded Core Test chapter describes the basic components of embedded core testing and describes some of the test techniques and architectures that are commonly used in parts of the industry today. When embedded cores are integrated into a whole chip-level architecture, then new DFT challenges are presented and reuse vectors must be managed. The test techniques and architectures described in this chapter allow core test architectures to co-exist with other chip-level test architectures to produce an efficient overall chip-level test architecture. This information is provided to illustrate the trade-offs involved with the various embedded core test processes so that the correct techniques, architectures, and proper vector management can be applied to meet cost-of-test, time-to-market, and quality-level goals.

5.1.2 Introduction to Embedded Core-Based Chip Testing

The ever-continuing march to smaller and faster silicon has opened the door to Ultra-Large Scale Integration (ULSI). A few years ago, a 1 million transistor, 50 MHz device was considered large and fast—now we are staring 100 million transistors and 1 GHz clock frequencies for commodity parts right in the face. The time-to-market (TTM) requirements and the short product lifetimes dictate that new chips have design cycles measured in months instead of years. Assembling a design team to "hand layout" transistors for these huge designs will never economically achieve the TTM goals—so design reuse, or cores, has become the solution to creating "massive functionality" in a relatively short period of time.

The ULSI-sized designs, system-on-a-chip (SOC) design methodology, the use of embedded cores, and the compression of the design cycle all individually present test problems. The creation of ULSI designs using embedded cores (see Figure 5-1) in a short period of time in one way is an easy test problem (if the cores are delivered with acceptable testability), but in another way is a severe test problem (in the creation of a chip-level test architecture, the mixing of multiple test strategies as delivered with each core, the management of the physical design, and managing the volume of test data).

To meet the aggressive TTM and quality requirements associated with today's products, some form of structured testability must be supported within each core and also at the chip level. Core-based designs cannot be made with "a new ad hoc" test architecture style with each design start—and a core cannot be just the previous generation chip with the pads knocked off. Cores must be made with reuse considered from the beginning, and a chip-level test architecture assembly process must be developed with consideration given to many cores, and their vectors, arriving from many sources (and possibly with many different types of individual testability supported). This implies that the "integrator" has a number of providers and has established a selection process and an acceptance criterion for incoming cores to be targeted for assembly into a chip-level device.

5.1.3 Reuse Cores

In order for system-on-a-chip to meet the TTM goals, the majority of the chip must be assembled from existing design elements with the "value added" or "design differentiation" logic elements being the components that may "set the schedule." From the testability point of view, the existing design elements, or cores, must be reusable not only in the design sense, but also in the test sense. Due consideration must be given to the type of testability (scan, BIST, functional, or pseudo-functional test), the amount of fault coverage, the type of fault coverage (stuck-at, delay, specification compliance), and the complexity of the test integration to the ultimate end user or integrator (the test modes, test interface, number of test signals, and test frequency). Also note that for "hard" (layout-only) or "non-mergeable" cores (cores that have inherent DFT), the vectors are also a deliverable that must also be thought of in the "reuse" sense: What vector format? what overall sizing? What individual pattern sets and individual pattern sizing? What data rate (frequency) when played back on the tester? What power consumption? What operation, timing, and edge set requirements are included that may restrict the usage to certain test platforms?

5.1.4 Chip Assembly Using Reuse Cores

The other major testability step required when dealing with a chip design comprised of a core, or multiple cores, is the overall chip test architecture. In the fantasy best-case scenario, all the cores have exactly the same test style (for example, Multiplexed D flip-flop full-scan), all the scan chains are of the same length (balanced), all the cores operate at the same test frequency, the package that the silicon will be placed within supports enough pins to bring all the core's scan chains directly to the pins for simultaneous operation, the operation of all reuse vectors simultaneously does not exceed the power limitations of the package or the silicon, and the delivered reuse vectors for all cores are compact, are fault-efficient, use low power, and will run on the lowest cost tester available. Unfortunately, the real-world case is more like: five cores from five different providers with five different test styles (MuxD scan, LSSD scan, logic BIST, direct bus access, and direct functional access); five different required chip-level test architectures to apply the reuse vectors; vastly different vector sizing and fault coverage ratings for the five different reuse vector sets; and each test vector set operates near the power limit of the package.

Given the real-world scenario, the integrator must assemble the chip in such a way as to make use of the different test architectures in a "chip-level" efficient manner to meet the overall test budget applied to the chip (many SOC designs are targeted for the general commodity market and are very cost-sensitive—and cost-of-test is becoming a dominating factor). In the case of multiple core integration, the test features of each of the cores now become part of the "functional" design of the overall chip (core test features should be considered as a delivered "function" of the core).

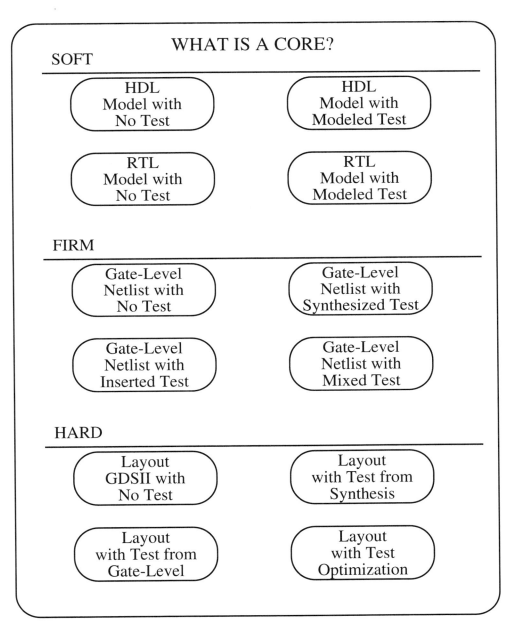

Figure 5-2 What Is a Core?

5.2 What Is a Core?

5.2.1 Defining Cores

There is no universally accepted definition as to what a core is (yet—see Figure 5-2). However, a few industry organizations are trying to address the definition and the problems inherent in core-based design—the IEEE with their P1500 Standard, and the private VSIA (Virtual Socket Interface Alliance) consortium.

Most engineers, if asked, will define a core as a complex piece of reusable design such as a microprocessor, video-processing unit, bus interface, or other pre-existing ASIC (application specific integrated circuit) logic (complex or not). Even a reusable JTAG TAP controller can be considered a core. Alternatively, some designers view hard stand-alone memory blocks as cores, and many others do not. Reusable design can be behavioral HDL or RTL, gate-level netlists (GLM), or physical layout (GDSII), and may be mergeable (can be mixed with other logic) or non-mergeable (the core is meant to be used as a stand-alone item without mixing with other logic).

5.2.2 The Core DFT and Test Problem

The adoption of a core-based design methodology includes the consideration of at least three different major test problems (and many minor ones). One test problem comes from the ability to create extremely large designs in a short period of time by reusing existing design elements (the system-on-a-chip design methodology); another problem involves the reuse of several cores within one design where they each may have existing but differing DFT strategies; and the third problem is having the ability to access and test complex cores when they are embedded.

The chips made by reusing existing designs and incorporating large amounts of reuse logic may include separated (existing) test strategies, which leads to the test methodology of testing the individual cores in isolation (the virtual test socket—being able to test each core individually in a self-contained manner). It may also be possible to use a unified or a single test methodology across the entire chip if the proprietary or intellectual property (IP) value of the core is not a concern (i.e., a merged design—simultaneous testing of all logic). The type of test strategies chosen for these designs may adversely impact other chip design budgets such as power, area, timing, packaging, and package pins. The trade-offs involving the type of test strategy chosen may also result in the adoption of the type of core (HDL, RTL, GLM, GDSII) and whether or not the core has pre-existing DFT logic.

5.2.3 Built-In DFT

One of the variables in core design, development, or selection is whether or not the core product has built-in DFT. Although there are different trade-offs for selecting cores that include (or don't include) DFT, the best guideline is that the organization that is going to fabricate (manufacture) and be responsible for testing the overall chip (and incur the cost-of-test) should define the DFT and Test strategy that is most cost effective for its business model. This approach removes the "over-the-wall" problem of testing a chip that may be made from a collection of diverse DFT strategies or testing a huge "system-on-a-chip" that may not include the correct DFT to undergo the test process economically.

5 • Embedded Core Test Fundamentals

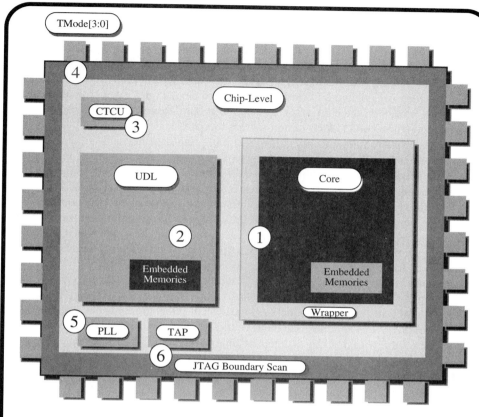

Figure 5-3   Chip Designed with Core

5.3 What is Core-Based Design?

5.3.1 Design of a Core-Based Chip

The core-based design methodology also does not have a universally accepted definition (see Figure 5-3). Some organizations believe that taking any pre-existing HDL, RTL, GLM, or GDSII and incorporating it within a design—with any design methodology it is considered core-based design. Other organizations consider it core-based design only if the core unit to be integrated is a large portion of the chip, and some organizations define core-based design to be only the reuse and integration of a HARD or non-mergeable cores.

The trade-offs and rationalizations involved with core-based design are also ill defined. Some organizations believe that core-based design is the only way that the aggressive time-to-market (TTM) windows for some chips can be met, whereas other organizations believe that the main purpose of core-based (reuse-based) design is to fill up the "white space" that results from the ever-shrinking process geometries and the minimum die size needed for economical package bonding. Some organizations believe that it is the only way to design "systems-on-a-chip" or ULSI devices.

In effect, all these aspects are true, but what philosophy an organization believes in may drive the type of business that is adopted. For example, an organization that wishes to make high-performance microprocessor engines and is not limited to cost issues may begin with the previous generation's microprocessor as an HDL unit and may build on it, add to it, or modify that HDL. However, an organization that is in the system integration business may begin with various hard (GDSII) cores and integrate them with other HDL and GLM cores to create a large (or small) system-on-a-chip in a short amount of time to meet a TTM requirement.

5.3.2 Core-Based Design Fundamentals

For this chapter, the simple model of the chip in Figure 5-3 will be used as the definition of core-based design. This chip model shows the units involved in core-based design that also affect the DFT or test issues of the overall chip.

As can be seen, the chip design includes its chip pin interface, at least one reuse core unit (which may be HDL, RTL, GLM, or GDSII), some non-core or user-defined chip logic (UDL), and some other features such as an IEEE 1149.1 (JTAG) architecture, an embedded clock source (VCO, PLL, Crystal Input, etc.), a chip-level test control unit, and embedded memory array cores (possibly within the logic core and within the non-core logic). Each of these logic units provides challenges to the DFT and Testability of the overall chip.

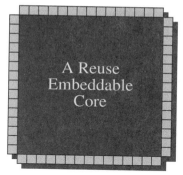

Figure 5-4 Reuse Core Deliverables

5.4 Reuse Core Deliverables

5.4.1 Embedded Core Deliverables

If a reuse core is to be embedded, then some support documentation and integration products must be delivered with it if the integration group is different from the core provider. Since a core may be a very complex device, such as a microprocessor, the receiving organization is highly unlikely to have an expert on the full extent of the core's functional features, test features, and the best way to use and integrate the core. As shown in Figure 5-4 any delivery of a core should, at the very least, include:

- the core (in HDL, GLM, or Layout Form)
- the data sheet that includes the core's features, operations, frequency of operation, power consumption, and the signal timing involved with the interface
- simulation models such as a "bus functional" model or an encrypted full-simulation description to allow software co-development (for example, an embedded operating system)
- other models such as interface timing and an ATPG interface (for example, an HDL, RTL, or gate-level description of a scannable test wrapper)
- an integration guide that describes how to use and connect the core in both the functional and the test sense
- the reuse vectors

From the test point of view, the delivered core should look like a reuse block that is going to be integrated as was intended by the core provider. So the integration guide should include an explanation of:

- how to connect and use any of the core's included test features
- how to integrate the core and apply quiescent defaults, if necessary, on test signals so that the test features are not active or destructive during any functional mode
- how to integrate the core so that the delivered "reuse" vectors can be applied as intended
- how to integrate the core and the reuse vectors so that chip-level test goals (such as support of Iddq testing) can be met
- how to connect clock signals to the core for test purposes that would minimize tester interface and on-chip clock skew problems
- how to incorporate the reuse test vectors into a test program

The delivered reuse test vectors should be the set that verifies the claims in the data sheet—vectors that verify the structure (operations), the frequency, the power consumption, and the signal timing specifications (for example, input-setup and output-valid). In addition, vectors should be delivered that can support chip-level test goals such as Iddq.

5 • Embedded Core Test Fundamentals

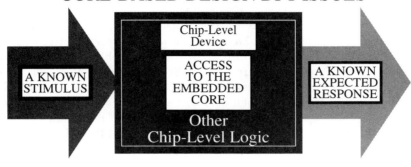

- If the Core is HARD—DFT must exist before delivery—how is access provided at the chip level?

- If the Core is HARD—and delivered with pre-generated vectors—how are vectors merged in the whole test program?

- If the Core is HARD—and part of the overall chip test environment—how is the core test scheduled?

- If the Core is HARD—and part of the overall chip test environment—what defaults are applied when not active?

- If the Core is HARD—what is the most economical and effective test mix—Scan? LBIST? MBIST? Functional?

- If the Core is SOFT—is the overall chip test environment developed as a Core and UDL or as a unified design?

- If the Core operates at a different frequency from the pin I/O or other chip logic—how does this affect DFT and Test?

Figure 5-5 Core DFT Issues

5.5 Core DFT Issues

5.5.1 Embedded Core-Based Design Test Issues

A core-based design methodology implies that a "whole" reuse logic block of some sort will be integrated or *embedded* within a chip. These logic blocks can be *soft* (HDL or RTL), *firm* (GLM), or *hard* (GDSII), and further, *mergeable* or *non-mergeable*. Different DFT and test issues exist if the embedded core is soft, firm, or hard, and mergeable or non-mergeable.

The most common implementations will most likely be the Non-Mergeable Hard Core (a layout macro with included test features) and the Mergeable and Non-Mergeable Soft Cores (synthesizable behavioral models without test, or with a defined test architecture, respectively—Figure 5-5). The layout macro is the easiest form to work with if the process library (FAB) is already selected. However, the synthesizable model is necessary if the ultimate design needs to be technology-independent (for example, the customer wants the design made both by the core provider and some second source vendor). Some reasons do exist for using a firm core (gate-level netlist)—the main reason being that the design can be mapped to a design library, with timing goals met, but still needs to remain open to different fabrication processes. There are different DFT and test issues for each type of core-based design.

5.5.1.1 Hard Core Test Issues

When a technology-dependent (process-mapped) layout macro is delivered to the chip integrator, then the DFT has already been accomplished and test logic exists in accordance to the strategy applied when the core was developed (scan, BIST, direct access)—this is the definition of a Non-Mergeable core (it already has the DFT provided). Delivered with the layout macro should be the specification, or data sheet; the integration guide; and the pre-existing reuse vectors. The data sheet should identify what operations the core conducts, what frequency range the core can be operated at, what power requirements are associated with the core, what the interface timing requirements are for the core signals (functional and test), and the clock tree delay and skew requirements.

The vector set delivered with the core should verify the claims in the data sheet (operations, frequency, signal timing, and power). The integration guide should include the following test-related topics: how these pre-existing reuse vectors are to be applied to the final design (the required test architecture on silicon), how the core should be configured to allow the application of the vectors, how the vectors should be incorporated into the overall test program, and any timing or special sequencing and edge set information related to the reuse vectors.

Other issues that need to be solved early in the core-based chip design schedule are:

- How is the core testing scheduled with respect to other chip units, or cores?
- When the core is not being actively tested, what test or functional mode is the core in, and what signal defaults are applied to the core (what is the safe mode)?
- How is access to be provided during core testing?
- Does the core need to change modes from functional to test, or vice-versa, on the fly for debug or failure analysis purposes?
- Is a test mode needed that has all the chip logic, including the core, active simultaneously—for example, a chip-wide iddq state?
- Should the chip test strategy match the core test strategy?

However, before the integration of a core in a core-based design methodology can be accomplished, the core itself must have met some basic DFT and reuse design rules. For example, during the core development, the test strategy for the hard core itself must be established. The decision to use functional access, scan, BIST, or some hybrid methodology depends on factors like:

- the number of test interface signals required
- the number of internal flip-flops in the core
- the complexity of the core unit
- the type of core unit (memory array, digital logic, analog circuitry)
- the overall cost target of core-based designs (system-on-a-chip, consumer electronics, high-performance products)
- the target integration environment (customer integrates core, core provider integrates core, or third party integrates core)

5.5.1.2   Soft and Firm Core Test Issues

The delivery of the synthesizable model for an embedded core design gives the most freedom for the overall chip design and integration. However, as a business decision, steps must be taken to protect the intellectual property (*IP*) value of the soft core by entering into legal partnerships, exercising non-disclosure agreements (NDAs), or by having the core provider being the core integrator.

The soft core can be provided in HDL or RTL, and it can be delivered with an existing test strategy included within the model (making it a Non-Mergeable core), or it can be delivered as a functional description-only with no test features included (making it a Mergeable core). Including test features in the behavioral model may be limiting to a soft core. For example, the core may be modeled as LSSD scan, and the overall chip design style may be Mux D-Flip-Flop—the eventual whole-chip design is then a complicated hybrid of the two different scan methodologies.

If no test features are included, one of the first design decisions required is whether or not the soft core will be treated and integrated like a hard core (Non-Mergeable—integrated and tested as a separate isolated unit), or whether all the HDL of the overall chip will be synthesized together and treated as one large database during test insertion (design flattening with Mergeable cores).

Note that all the decisions, concerns, and issues associated with the soft core are equally applicable to the firm core in that the netlist carries a similar IP risk—the netlist can be delivered with or without test—and the decision must be made to incorporate the netlist as an isolated unit or as an integral part of the whole chip. The only difference between the soft core and the firm core is that the netlist has already undergone the work to conduct synthesis (gate or design library mapping) and has met the necessary initial timing assessment.

Test insertion for a soft or firm core-based design depends on the overall chip testing strategy and is almost exactly the same as a whole chip development. This type of design is usually accomplished when the business model is to conduct "Turnkey" designs (a core provider also builds the whole chip, including cores, as the result of receiving a customer specification).

5.5.1.3 Frequency, Power, Timing, and Other Design Budget Issues

Several issues exist regarding the impact of hard, firm, and soft cores on the design budgets of the overall chip. Reuse may make it possible to incorporate a large amount of logic on a die (system-on-a-chip), and to place this die into a low cost plastic package; however, this methodology may also result in exceeding the power limitation imposed by the plastic package while the logic is undergoing manufacturing test (highly compressed scan vectors have been known to operate at three to five times the power consumption of functional operational vectors).

Another major consideration of whether to use hard, firm, or soft core design is the mixing and matching of logic or core units that have different frequency-of-operation requirements (different Fmax per design unit, or established clock relationships between certain design units). Several issues here deal with such concerns as delivering skew and delay managed clocks to the core, creating or using scan test architectures in such a way as to preserve the timing relationships between scan domains, and being able to deliver vector data from a tester at the rated frequency (for example, how can 100MHz test data be delivered from a 50MHz tester?). Some very basic concerns are "How to ensure that a core designed to run as some frequency still operates at that frequency limit after integration," and "What is the clock source for the core during functional operation and during test operation?" These issues involve the consideration of clock distribution, clock sourcing, clock skew and clock delay management, cell timing, overall core timing, the method of verification of timing after manufacturing, and sometimes, even the sizing of the power grid.

Finally, there are always the standard design issues (core-based or not): power distribution, numbers of powers and grounds, timing, area, chip-size, the size of the pin interface, route-ability, and other standard design concerns, and all these may be negatively affected by a core-based design methodology.

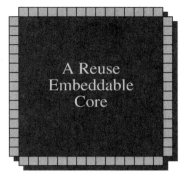

- DFT Drivers During Core Development

Target Market/Business—Turnkey versus Customer Design
Target Cost-Performance Profile—Low to High
Potential Packages—Plastic versus Ceramic
Potential Pin Counts

- Core Test Architectures and Interfaces

Direct Access—MUX Out Core Terminals
Add-On Test Wrapper—Virtual Test Socket
Interface Share-Wrapper—Scanned Registered Core I/O
At-Speed Scan or Logic Built-In Self-Test (LBIST)

- Design for Reuse Considerations

Dedicated Core Test Ports—Access via IC Pins
Reference Clocks—Test and Functional
Test Wrapper—Signal Reduction/No JTAG/No Bidi's
Virtual Test Socket—Vector Reuse

Figure 5-6 Core Development DFT Considerations

5.6 Development of a ReUsable Core

5.6.1 Embedded Core Considerations for DFT

The assessment of the impact of adopting a core-based design methodology must begin with the consideration of the target core itself.

When a design element is being considered for conversion to become a reusable core, or when a design element is going to be developed from scratch to become a reusable core, then one of the major concerns is what kind of DFT will be applied to achieve the required test goals of the core (independently of whatever final environment within which the core will be situated). The type of test strategy used, and therefore the DFT applied, depends mostly on the target business goals. Several possible test architectures can be applied to meet the overall goals, and some thought should be given to the ease of integration of the final core product (it is up to the core provider *not* to provide a great prickly porcupine that the integrator must grapple with). But no matter what DFT strategy is applied, the core must meet some fundamental "design for reuse" rules—and some thought must be given to "ease of integration" considerations (see Figure 5-6).

5.6.1.1 Business Considerations

When developing a core, the designing organization must think about the ultimate business application of the core. For example, if the core is going to be a very high-performance design unit to be embedded and integrated into very complicated high-performance chip environments, and the target chips are products designed and fabricated by the same organization that provided the core, then the core can be delivered as a soft core and it can be treated as HDL to be included with the other chip HDL. In this case, the DFT can be applied at the chip level and the core is just one more HDL unit within the chip. The same is true for small, low-performance chips made from a "parts bin" of peripheral cores—it may be easier to compile all of the HDL core units and *then* conduct a single synthesis and single chip-wide test insertion step.

However, if the core is going to be targeted for high-volume, low-cost, low-margin embedded designs, where the protection of the IP content is required or the provider and the integrator are in different organizations, then the transfer of a hard core may be the best business model. In the hard, or non-mergeable, core business, the DFT must be designed into the core prior to turning it into a GDSII file.

5.6.1.2 Test Architecture Choices

The test architecture choices depend on both the business considerations and the size or complexity of the core. If the core is small and simple and has a small signal interface, then it might be advisable not to apply any inherent DFT but just to recommend an integration method of multiplexing out all the core signals. The test methodology would be to access the core directly from the chip package pins and to apply an optimized set of functional vectors or structural test vectors generated by sequential ATPG.

If the business goal is to make a core that will go into a very low-cost market, then test cost may be a bigger issue. Also, if the core has a large signal interface, but is designed to go into inexpensive plastic packages (which may have small pin interfaces) then test interface reduction may be required. In these cases, a trade-off may have to be made to include a more efficient test

5 • Embedded Core Test Fundamentals

methodology such as scan to create an isolation test ring to reduce the effective test interface (a boundary scan that allows a few signals to access the functional interface in a serial manner), and to allow reuse test vectors to be made for the core prior to integration.

If the use of the core is even more cost compressed, or the core has high frequency but extremely low cost requirements, then an extreme test methodology such as Logic Built-in Self-Test (LBIST) might have to be used. This would require a larger investment in core DFT and test design and will result in extant core logic, but would allow the simplest (and most inexpensive) of testers to provide testing.

5.6.1.3 Design for Reuse DFT Considerations

The business requirements, cost requirements, and the complexity of the core will drive the type of DFT that needs to be applied to the chip. Certain conditions will result in always delivering an HDL core, and some conditions will result in the preference of delivering a hard layout macro-cell. The HDL core gives the most freedom for integration with respect to DFT (for example, scan insertion can be done at the core-level or at the chip-level), and the HARD Macro allows for the least (the existing DFT must be understood and integrated uniquely at the chip level). The key issue with delivering a consistently "integrateable" core, from the DFT point of view, is to "plan for reuse" in certain fundamental areas, since it is very unlikely that every possible integration scenario can be thought of ahead of time. Another way of saying this is "sweat the small stuff" or "mind the pennies"—if certain key areas are handled as a general rule, then all possible scenarios do not need to be enumerated.

For example, when a core I was involved with was first considered for core integration, the designers started finding impossibilities around every corner: "What if the customer puts it in a package with only 44 pins?" "What if they want to connect it to some internal tristate bus of unknown protocol?" "What if they connect the core to some internal clock generator like a PLL?" "What if we recommend that the core be operated at a certain frequency to meet performance needs, but the chip bus or pin interface will not operate at that same frequency (e.g., the bus operates at a slower frequency)?" The game at that point was for everyone to toss out a "what if," which is the same as enumerating integration scenarios. Although this was a good exercise, it can quickly, and falsely, lead to the assumption that "reuse can't be done." The real solution was to assess the business limitations and to address the key reuse areas.

A listing of the "reuse" areas of concern to concentrate on are (what to think about when the core is being developed):

- *The Test Interface*
  - **dedicated test control signals:** There needs to be "ease of application" of test modes—one of the easiest ways to enter and exit test modes is to provide direct input signals into combinational-only test mode logic.
  - **how many test signals:** In deep-submicron designs, timing, routing, and delay effects are very critical, and integrators demand minimal routes and borrowing or dedicating few pins to test—so fewer is better.
  - **using a scan wrapper:** The need is to provide a core that can support test-in-isolation and a reduction of functional interface (fewer is better).

- **using a test/scan wrapper:** Not only is test-in-isolation required (the ability to provide test and generate vectors independent of the integration), but a structure needs to exist that can assess the core's signal timing and frequency of operation.
- **JTAG:** A core should not support 1149.1 type testing with an on-core test access port (TAP) and TAP controller, because multiple TAP controllers on a chip is not a compliant design —however, supporting a boundary scan-like test ring is allowed.
- **no bidirectional ports:** Managing a net that can support a logic "Z" at the boundary of a core is a difficult problem in test control, test-in-isolation, and scan insertion.

- **The Test Clock(s)**
  - **bypass for test:** Reducing the complexity of testing the core, or applying the reuse vectors, and providing a low-cost test vector set is easier if the tester has direct control of the synchronizing clock during testing—this minimizes complexity in providing test if an on-core (or on-chip) PLL is supported and it allows testing if there are edge set and timing specification limitations.
  - **clock-out signal:** If no bypass clock will be provided, then helping the tester identify when to apply new input data and when to observe output data requires providing a signal from the core's clock tree.
  - **clock stopping or slowing:** The ability to stop the clock or slow the clock rate, is needed for Iddq or retention testing, or for power considerations (full-frequency operation may result in excessive power consumption or elevated thermal operation).
  - **at-speed clocking:** Full-frequency operation is required for power and frequency verification.
  - **managed clock domains:** Delay and skew management of the clock is necessary to enable simple scan insertion with minimum "shift races" and "hold time" problems—and to enable timing assessment (especially if there are multiple time domains).

- **The Test Strategies**
  - **simple scan:** A scan architecture allows for structural stuck-at coverage and will enable the use of ATPG to provide deterministic, fault efficient vectors.
  - **parallel scan chains:** Supporting several scan chains that operate in parallel provides a more cost-effective architecture by reducing test time (but this has to be weighed against the "fewer is better" test route preference).
  - **AC scan:** Supporting AC scan enables frequency and timing assessment.

- **at-speed scan:** Supporting at-speed operation of the scan chains enables reduction of test time and the ability to measure peak power during the shift operation.

- **multiplexor mode:** For a small, simple core with few pins, bringing the entire functional interface out to the edge of a chip for test is not unreasonable (but again, fewer is better and the vectors would be functional or structural sequential vectors from a sequential ATPG tool).

- **logic BIST:** This type of test strategy is best for high-performance cores with extreme cost of test limitations (pin limitations or access limitations).

- **memory BIST:** For cores with embedded memory arrays, the memory will be "doubly embedded" when the core is embedded.

- *The Integration Methods*

    - **isolated test:** Supporting the ability to separately test the core and any chip logic will provide for accountability, ease of debug, and ease of integration (true "plug and test" capability)—this can be a soft, firm, or hard core integration.

    - **merged test:** This integration method is based on soft or firm core integration where a chip-wide test architecture can be developed at one time.

    - **hybrid**: When multiple cores must be integrated and some are hard and some are soft/firm, or when some are mergeable and some are non-mergeable, then the overall chip integration will be a mixture of isolated and merged test.

- *The Test Requirements*

    - **structure**: The core must be verified for structure when embedded—this verification is accomplished most efficiently with scan.

    - **frequency and timing**: The core may need to be verified for operational frequency and signal timing—this verification can be accomplished with scan using delay fault models.

    - **Iddq current testing:** The core should support static design techniques and/or a quiescent state if the chip-level integration will require quiescent current or leakage measurements.

    - **characterization:** The chip-level design may be part of a new process development, or may require yield enhancement, and for engineering and failure analysis debug, logic observability and a memory bitmap output should be supported.

- *The Cost Concerns*
    - **vector management:** Core vectors must be integrated to be part of a chip test program, so the vectors can't be large and consume the majority of tester memory. The vectors should be compact and fault efficient (fault coverage per clock cycle should be high).
    - **vector compression:** The compression of core vectors will result in fewer vectors, but those vectors will have more activity/power consumption.
    - **test signal routes:** In deep-submicron designs, timing, routing, and delay effects are very critical, and integrators demand minimal routes and borrowing or dedicating few pins to test—so fewer is better. Many test signal routes can cost pins, area, power, and design schedule time.
    - **dedicated test pins:** Requiring dedicated test-only signals for the core becomes a chip package pin problem or cost.
    - **test power consumption**: Test activity is usually higher than functional mode, so test operation may use more power—the power grid of the core and the power requirements of the embedded core should reflect the test power rating.
    - **test parallelism:** Enabling the core to allow configurations where the core may be tested in parallel with other cores and chip-logic results in shorter test times but requires more power.
    - **test frequency:** At-speed scan results in less test time but also requires more power and a tester that supports the scan data rate.
    - **single edge set:** Supporting a single edge set for all core test modes allows for use of a simpler tester and for development of a simpler test program.
    - **time-to-market:** The core should be developed with a thought to the most simple test integration possible—this requires evaluating the test strategies and test signals for their impact on the integration time involved with the DFT features and the merging of reuse vectors into the chip-level test program.

Figure 5-7   DFT Core Interface Considerations

5.7 DFT Interface Considerations—Test Signals

5.7.1 Embedded Core Interface Considerations for DFT—Test Signals

The number of signals on an embeddable core and the method of integration are important considerations. The very first consideration for DFT is "how many signals will be available to test the core when it is embedded?" If the business model allows for the core to be embedded in a very low-cost plastic package, then the number of functional signals and the number of test signals becomes a major issue. For example, a hard core with 60 signals, when placed within a 44 pin package, cannot have all its pins multiplexed to the edge of the chip package for functional access unless some design effort is applied to make a pin interface cache to time-multiplex the pin data—and note that this type of test design will not allow the core to process code on a normal cycle-by-cycle basis, a fact that may negatively affect the structural or timing test coverage involved with using the functional vectors.

Also note that this is a "what if" scenario, and that, in reality, during the core design and development it is not known what the real application or specific implementation that a customer may request would be. To this end, the core producer must plan ahead of time to design a core that will meet the needs of most customers (defined by the business case).

For DFT and Test interface considerations, the items that need to be addressed are summarized in Figure 5-7 and detailed as follows:

How many "dedicated" test pins are needed in the core interface? This question means, "How many signals must be brought to the chip pin interface that require dedicated test-only pins?" When business, cost, and design considerations are factored into this question, the answer is "as few as possible." If no dedicated pins are needed to enable core test modes (enable the test logic and signals), then the core test logic signals must be directly active, and the safety defaults (the configuration of these signals when not in test mode) for these must be placed in the UDL area. If at least one dedicated pin is needed, then the safety defaults for the test signals may be provided within the core. For example, supporting two dedicated test mode signals may allow four different test modes where the "00" case would allow functional mode with all test inputs gated to receive constant logic 0s; a "01" case would open the scan gates and allow scan data and the scan enable to be applied; a "10" case would open the MBIST gates and allow the memory BIST signals to be operated; and finally, a "11" case may enable a quiet static mode to allow Iddq testing.

How many test signals need to be shared with the chip interface? This question means, "How many signals must be brought to the chip pin interface, but can share the functional pins?" Similarly, the answer to this one is also "as few as possible." In some cases, the answer to this question can limit the business applications of the core. For example, a core with a set of 32 scan chains and a set of 2 wrapper scan chains and 3 scan enables (core_SE, wrapper_inputs_SE, and wrapper_outputs_SE), requires 64 + 4 + 3 = 71 data type pins. This fact limits the core to only packages that have more than 71 data pins (plus powers and grounds). If the business case is to also support core integration into 68 pin plastic packages, then this scan architecture is not correct—and a trade-off will have to be evaluated for test time versus the number of scan chains (fewer scan chains will result in longer applied test clocking).

What are the frequency requirements of the test signals? This question means "at what frequency/data-rate are the core test signals designed to operate, and will these signals work with any package pin interface that the core may be embedded within?" (Even more specifically, "Will the testers available to test these parts be able to handle the test data rate requirement?") For example, if the core is designed to operate at a higher frequency than the chip interface (e.g., 2:1 ratio), does the chip interface have to support pins that operate at the higher core frequency just for test, or can the core test signals operate at the lower chip package frequency?

What is the planned integration method for these test signals? The answer to this question is also, "As simple as possible." This means that the most straightforward method of bringing test signals to the edge of the chip and borrowing functional pins, while preserving pin timing (edge sets), is preferred. For example, if the core is made with a dedicated test pin to enable scan mode, and there are 16 scan chains, with 1 scan enable, and these scan chains have a safety gate on all inputs—then the integration is as simple as bringing 16 scan input wires to any 16 chip input pins, bringing the 16 scan output wires to any 16 chip output signals through a multiplexor, bringing the dedicated scan mode signal to a package pin, and connecting the scan mode signal to the select line of the multiplexors. When the core scan mode package pin is asserted, the data applied to the selected 16 scan inputs pins and the selected scan enable pin, pass through the safety gate and apply scan testing to the core. When the core scan mode package pin is de-asserted, the safety gates place a constant logic 0 on the scan inputs and scan enable input, regardless of what the package pin data is, keeping the core's scan architecture in a quiescent mode.

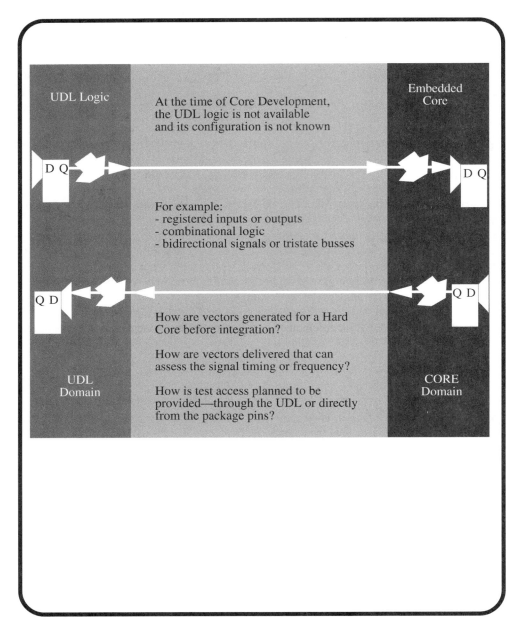

Figure 5-8 DFT Core Interface Concerns

5.8 Core DFT Interface Concerns—Test Access

5.8.1 Test Access to the Core Interface

One of the most fundamental problems faced when designing a core is determining the method of achieving access to the core, for test purposes, after it is embedded (see Figure 5-8). This is an integration engineering process, because the core interface does not connect to anything during core development. The core design team must determine whether the test will be delivered by the natural functional pathways from the other chip logic, or whether the direct test access will enter by alternate pathways.

Delivering test access by using the natural functional interface cannot take into account the timing involved with delivering vectors to, or observing vectors from, the core interface, since the timing of the complete path is not known at the time of core development. This implies that the timing vectors (as well as the structural vectors) must be created after the core is integrated, and late in the design when the overall chip design and timing are known. Testing after integration is really a feasible methodology only if the overall chip design methodology is based on using a mergeable soft core, which will be test inserted and will undergo ATPG after the chip-level integration process (at least then the vectors can be delivered in a timely fashion).

If the core will be delivered as a hard core, then the test access method needs to be defined and developed at the time of the core design, not at the time of integration. This is because the hard core must have the DFT already included at the time of deployment, and the core designers must have placed the DFT within the core during the design phase. If the hard core is delivered with pre-generated reuse vectors, then the test access method is the method that will allow the reuse vectors to be applied as generated. The possible methods for hard, or non-mergeable, core integration are:

Do Nothing Integrate the core to be functionally tested—this has a high vector cost and a possible fault coverage impact. Pre-existing reuse vectors cannot be delivered with the core—they must be developed by the integrator with the simulation model of the core (which may be encrypted—so how are they fault graded?). A time-to-volume issue may arise if developing vectors for coverage of the embedded core takes months to achieve a qualification level.

Direct Access Integrate the core by bringing all the core signals to the chip package pin interface—this has a high routing cost and a possible fault coverage impact. Vectors can be delivered with the core, but not for core interface timing, since the final core interface will not be established and timed until it is embedded within the chip.

Slow Boundary Scan (JTAG) The integration method is to provide functional vectors, but through some slow serial boundary scan elements—this method reduces the number of chip-level routes, but requires converting functional vectors to serial scan vectors, and does not address the timing assessment requirements (JTAG boundary scan is based on serially shifting in the value of the pin interface, and then applying or "updating" that state as one action). Note that a core may support only a boundary scan ring, not a TAP controller, under the IEEE 1149.1 Standard—currently, a compliant chip cannot have two TAP controllers with included bypass registers.

Built-In Self-Test The integration method is to use embedded LFSRs with the core to generate vectors and compress (signature analyze) the response—this has a design and overhead cost, may have a possible fault coverage impact (random vectors are not as fault coverage-efficient as deterministic vectors), may have a test time impact (it takes more random vectors to get equivalent coverage), and does not really address the timing assessment problem. BIST may also affect the design schedule as the LFSR's must be designed and chosen for their coverage, so experiments and fault simulation may be required to assess the goodness of the LFSRs to be used.

Stored Pattern Test The integration method is to deliver a set of vectors with the core, and to require them to be placed in on-chip memory. This method has a high area cost if it requires a dedicated memory (such as a ROM) or a high engineering design schedule cost if it requires making access to a system memory (such as an SRAM or EEPROM). Also the vectors may have fault coverage gaps if the test application architecture is "specialized" and does not match a "functional interface."

At-Speed Scan The integration method is to apply some number of parallel scan chains to both the core's internal and the signal interface logic. This allows for a reduced test interface (fewer global routes—just the scan connections) and also allows for "in-isolation" timing assessment (the virtual test socket). In this case, reuse vectors can be generated by ATPG during the core design and can be delivered with the core.

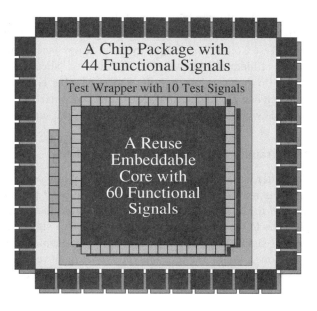

- Core DFT Interface Considerations

Wrapper for interface signal reduction

Wrapper for frequency assessment

Wrapper as frequency boundary

Wrapper as a virtual test socket (for ATPG)

Note: bidirectional functional signals can't cross the boundary if wrapper or scan

Figure 5-9 DFT Core Interface Considerations

5.9 DFT Interface Concerns—Test Wrappers

Although it is the addition of more logic, one of the most effective reuse methods that can be applied to a potential core design is to support a design unit that has been called a: "Test Ring," "Test Wrapper," "Test Collar," or "Boundary Scan Ring" (see Figure 5-9). These units can be an additional layer of hierarchy known as an "add-on" or "slice wrapper," or they can be "designed-in" and integral to the core design itself. A slice wrapper is generally thought of as a group of logic elements that can be "laid across" the natural core interface, whereas the integral, or built-in, wrapper is made with the core interface logic, and is not separable.

5.9.1 The Test Wrapper as a Signal reduction Element

The wrapper, no matter what form, can be used for several purposes. If designed correctly, the wrapper can reduce the number of signals required to fully test the core after it has been embedded. For example, if the core has 130 functional signals but these are tied together into a 130-bit scan chain, then the I/O map can be reduced to 3 interface signals (a scan input, a scan output, and a scan control signal, scan enable). This, however, enables testing of the I/O only and, if nothing else is done, would require applying functional vectors converted to serial scanned-in data or the application of a sequential ATPG tool to generate core vectors through the serial access.

If, in conjunction to the I/O scan chain, the internal logic is also tested by bringing the core scan chains to the wrapper boundary, then scan can be used to apply the internal state and the I/O map simultaneously. For example, balancing the multiple internal core scan chains to be 130 bits deep to match the I/O map scan chain may require 16 scan chains. So now the whole interface has been reduced to 36 signals (16 parallel core scan inputs, 16 parallel core scan outputs, 1 wrapper scan input, 1 wrapper scan output, and the 2 scan enable signals, wrapper_SE and parallel_SE). During test mode, these are all the signals that are required to test the core, so instead of dealing with a test integration of 130 signals, only 36 signals must be routed to the package interface.

5.9.2 The Test Wrapper as a Frequency Interface

If the core is designed to operate at some FMax frequency and it is also designed to be embedded in a design that will surround it with logic that operates at a different frequency, then the best course of action is to design the core interface at the test wrapper to also be the frequency boundary. Usually this is done in one of several ways: the core and wrapper have a different clock domain from the other chip logic; or there is an interface term in the wrapper that informs the core when incoming and outgoing data is valid; or there is a known cycle ratio for incoming and outgoing data to be valid (e.g., the bus-to-core transfer is 2:1 or 3:1). If scan is to be applied to allow at-speed frequency assessment, then the ability to preserve the timing during the scan sample cycle is important. This means that the wrapper must accommodate the control signal or clock signal, used for time domain separation, as a valid test signal.

5.9.3 The Test Wrapper as a Virtual Test Socket

The primary purpose of a test wrapper is to allow "test-in-isolation" to occur. This means that the full spectrum of vectors can be generated for the core prior to delivery of the core (and in the absence of any other logic), and the wrapper will allow these vectors to be applied even if the rest of the chip is DOA (inoperable). This type of methodology allows test accountability between the chip logics for debug purposes: If the chip fails, is the cause the core or the UDL logic? Who does the debug?

There are basically two types of test wrappers—the "add-on" or "slice" test wrapper which is independent of the core and can be added on after the core is designed; and the "shared," "merged," or "registered" wrapper, which is an integral part of the core and is made from the core's own interface registers.

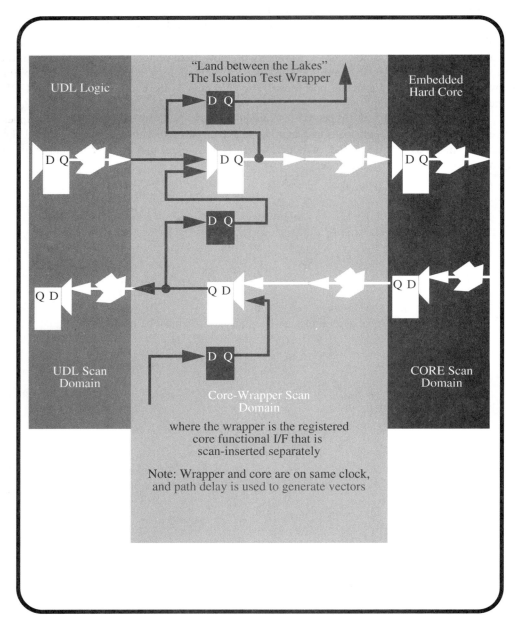

Figure 5-10   Registered Isolation Test Wrapper

5.10 The Registered Isolation Test Wrapper

The easiest type of test wrapper to consider is the registered shared-wrapper. This test wrapper architecture is called the shared-wrapper, because it is really the core's natural interface logic. Some design organizations believe that creating a core device with a completely registered natural interface is a large performance penalty, but in reality, it is the most effective reuse technique. This test wrapper is made by conducting a directed scan insertion of only those sequential elements that are in the natural interface and placing them in their own scan chain (or chains) with their own scan enable control and with no other core sequential elements (see Figure 5-10). When an input cell captures data from the UDL, it is capturing data into the "real" data pathway, and when an output cell drives data into the UDL, it is driving from the "real" data pathway. Note that the core should still be scan-inserted, but as a separate action that creates the core scan domains separate from the interface scan domain.

To make this type of test wrapper into a "virtual test socket" for all test applications (stuck-at scan and AC scan), the natural bits must be interlaced with independent shift bits. These are "dummy" non-scan bits that are interlaced into the boundary scan architecture and that allow "vector pairs" to be launched from the test wrapper.

The wrapper is a device that is used by both the core and the UDL, so a gate-level non-encrypted description of it must be delivered to the integrator so that UDL timing analysis and vector generation can be done. This requirement is one of the differences between the shared-wrapper and the "add-on" wrapper—delivering a gate-level description of the "add-on" wrapper is straightforward since a behavioral or gate-level description must be made as a separate action for the original purpose of vector generation for the core. However, extra work must be done to extract a "non-encrypted" shared-wrapper gate-level description, since it is inseparably embedded within the core (and the key is to keep the internal configuration of the core proprietary so that only the wrapper should be delivered). To conduct an accurate timing analysis, the delivered test wrapper should exactly match the "real" test wrapper with the same elements, buffering, and timing definitions.

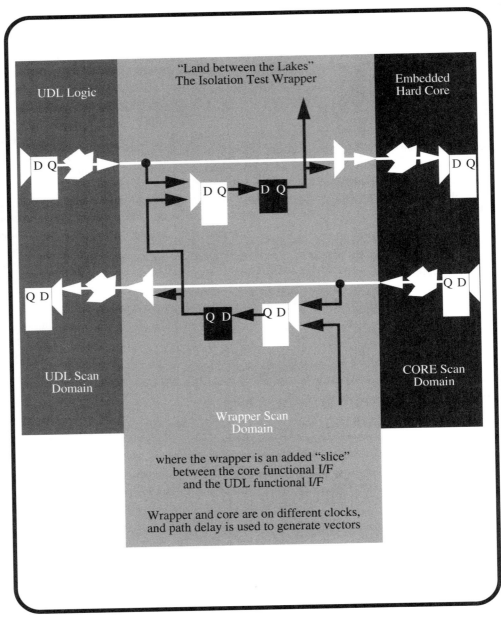

Figure 5-11 Slice Isolation Test Wrapper

5.11 The Slice Isolation Test Wrapper

The "add-on," "slice," or "partition-cell" test wrapper is implemented by stringing together partition cell elements to form a boundary scan (see Figure 5-11). This test wrapper is different from the IEEE 1149.1 (JTAG) type boundary scan element in that both elements are on the scan shift chain and are operated with "direct" tester control by the same clock-edge, not by a TAP controller or TAP state machine.

In JTAG, the first element is known as the "capture and shift" cell, and the second element is known as the "pre-load" or "update" cell. Only the first cell is part of the scan chain to prevent the random shift data from toggling the downstream logic. The at-speed "Partition-Cell (PCell)" test wrapper is used only in test mode (JTAG's boundary scan is designed to also allow transparent operation during functional mode), so toggling the downstream logic is not a concern—by moving the scanout connection from the first element to the second element, the scan chain allows "vector pairs" to be launched (note: the JTAG cell also clocks the capture cell on the positive edge of TCK while clocking the update cell on the negative edge of TCK—whereas the PCell element clocks both elements on a positive edge clock). This configuration is what enables the core or the UDL to be tested for signal timing specifications. A path delay vector generation (ATPG) tool can be used to automatically create the vectors for input-setup and clock-to-out or output-valid vectors.

The PCell test wrapper can be thought of as a removable "slice" or as a level of hierarchy that encompasses the core and is easier to deliver to the customer, since it is separated from the core.

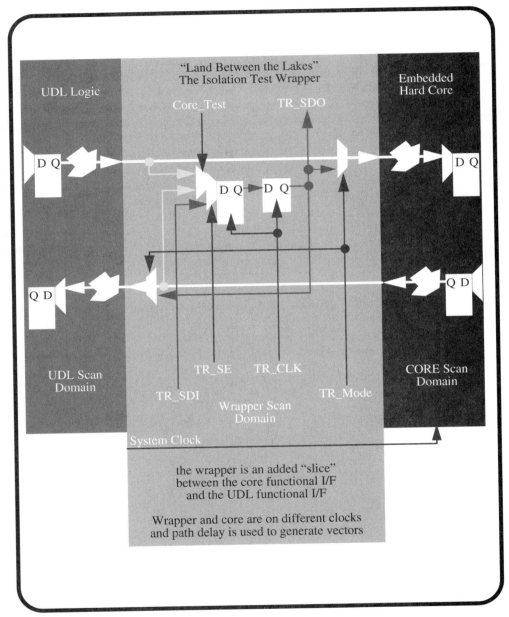

Figure 5-12 Slice Isolation Test Wrapper Cell

5.12 The Isolation Test Wrapper—Slice Cell

The isolation wrapper cell shown in Figure 5-12 is one example of a test wrapper element that can be used to enable "test-in-isolation" as well as "at-speed" interface signal assessment. This cell has two scan elements to allow independent transitions, or vector pairs, to be associated with each core signal (input or output).

This particular cell shown in Figure 5-12 shares the two wrapper scan bits between two multiplexors, so a single cell may handle one core input and one core output. This configuration is an optimization that saves applying two flip-flops and a multiplexor to every core signal (and note that the second element does not need to be a scan element, since it is just an independent non-functional shift bit). However, this optimization does cost another integration signal, Core_Test, to indicate the direction of testing (testing the UDL or testing the core).

As can be seen, this cell requires a test ring Scan input (TR_SDI), Scan output (TR_SDO), Scan Enable (TR_SE), and a Scan Test Mode signal (TR_Mode). Note that this particular cell has a different test clock (TR_CLK) than the core system clock (or the UDL system clock). This is because this cell is being used as a slice wrapper that is placed across the core-UDL interface signals. The timing assessment using this type of wrapper is more complicated, since the test must be done in halves: a test must be generated from the UDL to the wrapper with the proper timing as reported by a timing analysis, and a test must be generated from the core to the wrapper with the proper timing as reported by a timing analysis (or a defined specification can be used)—the timing is verified by adding the timing involved with the two tests. Note that the individual vectors for the core can be generated with just the wrapper and the core, and the vectors for the UDL can be generated with just the wrapper and the UDL—these vectors will be generated at different times, so the timing of the interface paths may not be verified until both sets of vectors exist.

Occasionally, a core designer may need to consider a cell that may also be used by a chip-level JTAG controller. If this is the case, then this cell may be converted to a standard JTAG cell by application of two multiplexors. One multiplexor would select the scan data out point as being from either the capture cell (JTAG) or the update cell (at-speed). The other multiplexor would apply the clock to the update cell as a negative edge version of TR_CLK (JTAG) or as the positive version of TR_CLK (at-speed).

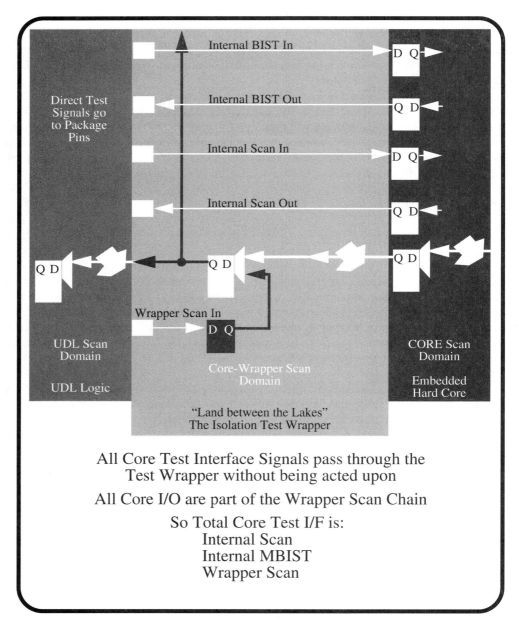

Figure 5-13 Core DFT Connections through the Test Wrapper

5.13 The Isolation Test Wrapper—Core DFT Interface

The test wrapper can be used to reduce the size of the core's test integration interface. This reduction can be done by effectively wrapping all the functional I/O into a set of scan chains, and to bring out only the internal core direct test signals (see Figure 5-13). In a typical reuse core, these signals would be:

- Internal Core Parallel Scan Inputs and Outputs
- Internal Core Parallel Scan Enable Input (per clock domain)
- Wrapper Parallel Scan Inputs and Outputs
- Wrapper Parallel Scan Enable Input(s)
- Wrapper Parallel Scan Mode or Direction Control Input
- Internal Core Memory BIST Invoke Input
- Internal Core Memory BIST Retention Input
- Internal Core Memory BIST BitMap Input
- Internal Core Memory BIST Fail Output
- Internal Core Memory BIST Done Output
- Internal Core Memory BIST BitMap Output

The term "Parallel Scan" means that more than one scan chain may be brought out to the edge of the core (or wrapper).

This type of wrapper configuration can turn a core with 250+ signals into an easy test integration that requires only routing a minimum number of signals to the pin interface (for example, 13). If multiple parallel scan chains are supported, then the interface impact is 2 signals per scan chain plus the scan enable (e.g., one scan chain is 3 signals, two scan chains are 5 signals, three scan chains are 7 signals, and so on). Note that at least one scan chain and one scan enable should be provided for each different clock domain as well.

The key point here is that the internal core test control and data signals are brought through the test wrapper directly without any interaction with the test wrapper. Similarly, the test wrapper control and data signals are also brought out this way. These are the only signals that need to be connected to the package pins from the core (unless some core operation signals have equal or higher priority than the test signals, for example, a global core asynchronous reset or signals to a clock-gating structure).

For timing purposes, the integration of these test signals should also be easy in timing. As Figure 5-13 illustrates, all these test signals are registered in the core/wrapper. This allows the most leeway in routing the signal wires directly to the edge of the chip during integration. Since vectors will be delivered with the core, any additional registration should not be done at the chip level, or else the vectors may have to be changed to add extra cycles after delivery.

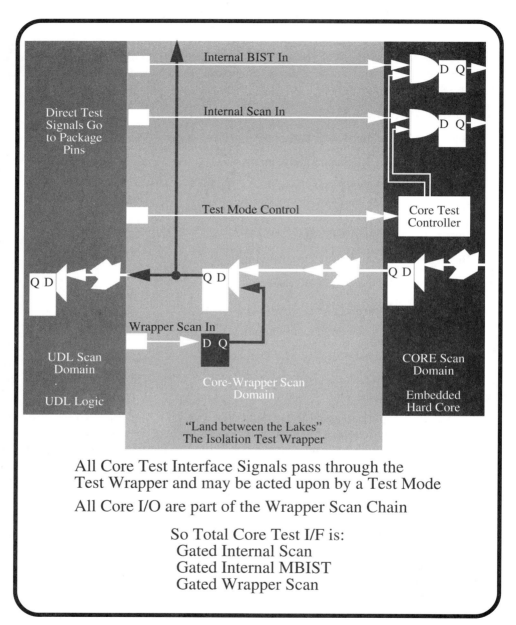

Figure 5-14 Core DFT Connections with Test Mode Gating

5.14 Core Test Mode Default Values

5.14.1 Internal versus External Test Quiescence Defaults Application

A decision must be made during core development concerning where the test logic defaults, or safety logics, are applied. Either the core can have a set of test modes that are applied through dedicated test mode control signals and the test modes would enable or disable the incoming test signals, or the test signals can be left active at the core/wrapper boundary and the defaults can be applied at the chip level during integration. The trade-offs involved are:

A Self-Defaulting Core requires dedicated test control inputs at the core boundary and must incorporate an internal test controller (see Figure 5-14). The test controller must be designed to allow all possible legal test modes and combinations of modes. For example, if more than one scan mode is used where one scan mode enables the memory arrays for burn-in scan operation and another scan mode disables memory writes during full at-speed production test scan operation, then one way to differentiate the two would be through a scan mode selection built into the core. At the chip level, the integrator must create selection logic to address the dedicated test mode select inputs on the core.

This type of design is generally considered to be safer in that the test architecture can be used only the way it was intended. Test signals can be used only when the test mode select logic enables them. However, the core designers must establish and implement "all the legal cases," without knowing every possible integration situation.

A Non-Defaulting Core does not require any test selection pins, but the operation of the various test signals in an incorrect manner may be possible. For example, if the memory arrays are tested by BIST and the BIST controller is scanned, then the BIST and the scan modes must not be operated at the same time. However, it is up to the core integrator to create the chip-level selection logic and to provide the default values for the test logic when the test logic should be quiescent. Note that the method for this type of core to select between two types of scan modes would be to bring special control signals to the edge of the core—for the example given in the Self-Defaulting Core, the signal needed would be a "memory write inhibit" signal to select a quiet memory versus an active memory during test.

The trade-off between the two methods depends mostly on how many test signals are needed within the core to exercise the core logic. If there is a wealth of signals, then bringing them all out would be more painful to the integrator than decoding a few test mode signals. However, if only the scan and BIST data inputs and control signals are needed, then there is no reason to add extra signals to separate the modes.

In any case, the integrator needs to be informed on how to treat all the test signals that are evident at the edge of the core/wrapper boundary.

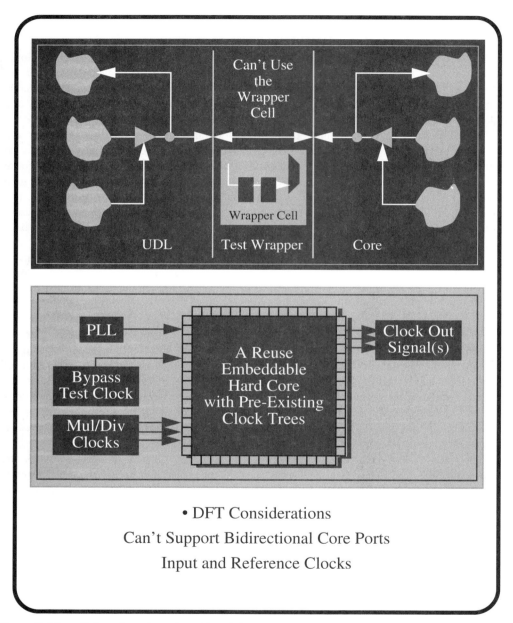

Figure 5-15 Other Core Interface Signal Concerns

5.15 DFT Interface Wrapper Concerns

Another set of concerns for the development of a core interface, with reuse considerations, is the directional nature of the signals in the core's natural or functional interface, and the clock or clocks used for testing (see Figure 5-15).

5.15.1 Lack of Bidirectional Signals

A significant problem with core-based design has to do with the nature of the core's signal interface. The core design organization naturally thinks in terms of minimizing the interface. This usually leads to thoughts of using bidirectional signals in the core interface and results in severe DFT, design methodology, and integration problems. For example, for a bidirectional signal to pass through the test wrapper, all the tristate control signals, in both the core and the UDL, that pertain to the tristatable signal must be controllable from the test wrapper. This requirement is difficult at the time of core development, because the UDL tristate enable signals are not available yet. Vectors cannot be generated for the core if the direction of the bus cannot be determined. If vectors are developed in the absence of the UDL and with some assumptions, then when the vectors are applied, some contention may exist. This type of situation may place some unreasonable restrictions on the UDL during core integration (for example, requiring that the UDL designer make an "all tristate" safe mode that is active anytime the core is being tested).

Another issue to consider is the design problem of "what strength driver" is the tristate device. The core designer must select the drive strength of a bus driver without knowing the final architecture of the bus—tristatable wires can't be buffered for drive strength.

5.15.2 Test Clock Source Considerations

Another test consideration is "not knowing the test clock source" at the time of core development. When integrated, the core may receive its functional clock from an on-chip clock source (e.g., a PLL), or it may receive its functional clock directly from a package pin or from a package pin or PLL and passed through some multiply or divide clock generation logic (in some cases, the core itself may be self-clocking and come with an on-core PLL). If the core is delivered as a hard core, then the internal clock trees will already be in place and the core should have been skew- and delay-managed when going through place-and-route. The recommendation should be made that the core be tested by providing a bypass clock source (to allow the tester to test the core from a clock-in point of view where the tester provides the synchronizing clock that it uses as a reference to schedule new applied data and to observe output data).

Sometimes, however, design budgets and considerations may lead the integrator to provide only the clock input from a non-bypass or non-tester source such as the on-chip PLL. In this case, all the testing done must be referenced to a clock-out signal provided by the core from the core's clock tree, and a procedure known as an edge search will have to be conducted by the tester. To this end, either a bypass test clock will have to be stipulated as the only way to apply the pre-existing core vectors, or the core will have to be made with a clock-out signal and a set of time-adjusted vectors will have to be delivered that can be referenced from the clock-out signal.

If the overall chip will undergo Iddq or any current measurement testing, then a direct bypass clock signal from the tester is required, or else the clock source to the core (and other chip logic) must be gated. This is because most on-chip clock sources have a "lock" time associated with them (a number of clocks before the clock signal is stable in frequency and edge rate), and stopping the clock to conduct retention or quiescent current testing may result in an unstable clock after the clock pause. Note that gating any clock not provided directly from the tester, to the core or any embedded memory arrays may also be necessary as a method to manage power consumption during the test process as well.

5 • Embedded Core Test Fundamentals

- Core DFT Frequency Considerations

Wrapper for frequency boundary

Test signals designed for low frequency

Package interface designed for high frequency

Wrapper as a multi-frequency ATPG test socket

Note: functional high/low frequency signals can cross the wrapper—the test I/F is the concern

Figure 5-16 DFT Core Interface Frequency Considerations

5.16 DFT Interface Concerns—Test Frequency

5.16.1 Embedded Core Interface Concerns for DFT—Test Frequency

One of the more difficult problems to consider when designing a reuse core in the absence of the chip package and other chip logic is the frequency of operation of the core's test logic and test interface (see Figure 5-16). Although the core only needs to be tested at the frequency of the final chip design (the functional frequency or frequencies that the core will be operated at after it is embedded and integrated), this is not known when the core is being designed. So the core designer must plan for the worst-case scenario—that the core will operate at its Fmax rated frequency and that the chip non-core logic and chip interface may operate at some frequency that is much lower (or higher). This is an especially severe issue if the vectors are delivered with the core (since they may convey the frequency and timing within the delivered format).

Another item of information that may not be available at the time of core development is the operating limit of the target tester (e.g., if the core can be mapped to several fabrication facilities, then the test floor to test the final chip may not be known). This means that the frequency target of the vectors delivered with the core must allow for "any" available tester, or the tester "class" must be specified with the delivered core and vectors.

5.16.2 Solving the Frequency Problem

The easiest way to solve frequency problems is to generate all core vectors to operate at the lowest possible frequency—the lowest frequency limit of both the chip interface and the known family of testers—this, however, may compromise the ability to test for AC goals. Another method is to fully specify how the core is to be integrated and tested, and this specification should include the package pin frequency rating for the test integration and the target test platform.

If no effort will be made to limit the business applications of the core (you'd be surprised how often this situation happens), then the delivered vectors and test mode operation must take into account the lowest possible frequency. This generally requires the scan vectors to be delivered to the part at a reasonable data rate, usually at 1 to 10 MHz. However, for AC verification, the scan sample(s) is required to be applied at the rated frequency of operation of the core. The method to allow this consists of playing a duty-cycle trick of changing the duty-cycle of the clock waveform on the sample cycle or cycles so that the sequential updates occur at the proper interval—sometimes this method requires other tricks such as scanning in the core vectors based on a slow tester-provided clock, and then to conduct the sample using the on-chip PLL.

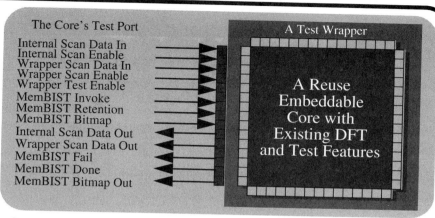

- Core DFT Goals and Features
  Embedded Memory Test by MBIST
  - Few Signals—High Coverage—Less Test Time
  - Bitmap Characterization Support
  Structure by Stuck-At Scan
  - High Coverage—Fewer Vectors—Ease of Application
  Frequency by At-Speed Scan (Path & Transition Delay)
  - Deterministic—Fewer Vectors—Ease of Application
  Reuse of Core Patterns Independent of Integration
  Test Insulation from Customer Logic
  Embedded Core I/O Timing Specifications with Wrapper
  Minimize Test Logic Area Impact
  Minimize Test Logic Performance Penalty
  DFT Scannability Logic
    Full-Scan Single-edge Triggered MUX DFF
      Tristate Busses - Contention/Float Prevention
    Negedge Inputs and Outputs
  Iddq—No Active Logic and Clock Stop Support

Figure 5-17   A Reuse Embedded Core's DFT Features

5.17 Core DFT Development

If the core is developed for reuse and the core is to be a hard core or a non-mergeable soft or firm core, then several DFT and test goals and features need to be addressed at the time the core is created (see Figure 5-17). As was mentioned earlier, the business and cost goals of the core, and the fabrication cost requirements will drive the type DFT applied. However, for illustrative purposes, an optimal method with the trade-offs explained will now be described. This described method will include internal at-speed AC scan, at-speed AC I/O wrapper scan, and memory built-in self-test. All of these taken together result in a minimal test interface of at least 10 signals that can be extended up to the number of whatever the business needs can support. However, as far as integrators are concerned, one of the most important issues in using a core, especially in deep-submicron (or nanometer) design, is the number of global signals that must be routed from the core to the chip pin interface to enable testing—the fewer the better.

5.17.1 Internal Parallel Scan

The core itself, if it is of a size and complexity to support an aggressive test methodology, should support at-speed parallel AC scan. This is the use of several scan chains that allows full testing (controllability and observability) of the internal logic of the core and that passes directly through the test wrapper to allow direct testing of the core. The smallest possible scan test interface is one scan input, one scan output, and one scan enable. However, as many scan chains as can be supported by business and cost considerations, to get the "shift depth" down to a reasonable number while still maintaining a reasonable interface, should be supported.

For example, if a core design is comprised of 1000 flip-flops and 5 scan chains would result in 200 bits (clock cycles) per scan vector load, then the scan interface would consist of 11 signals (5 scan inputs, 5 scan outputs, and 1 scan enable). This same design with 10 scan chains would result in a shift depth of 100 bits per scan vector, but at a cost of 21 signals. The trade-off is test data depth on the tester versus the number of scan channels. Note that test data depth does not always set the test time; for example, if the scan tests are applied at the rated frequency (at-speed), then the test time impact may be minimal (e.g., 2 MB of test data applied at 100MHz results in a test application time of 20 milliseconds—clearly the issue here is memory depth, not test time).

The at-speed parallel scan architecture will allow deterministic AC and DC vectors to be created by ATPG. With the tools available currently, the scan vectors can be "clock-cycle" compressed, allowing them to be rated for their different fault coverages and their vector time or vector space efficiency. Scan vectors can also be generated with a limited set of tester edge-sets (hopefully only one). This leads to the ability to make simple test program components and to apply them on less complex testers. Single edge set testing is accomplished by ensuring that each test signal is referenced from only one clock, uses a single tester format such as "non-return" (NR), and has only one worst-case input-setup point and output-valid strobe defined.

5.17.2 Wrapper Parallel Scan

The test wrapper associated with a full at-speed AC scan core should also support at-speed AC scan to supply the core I/O logic values at the same time the core is being scanned. To ensure the most efficient use of vector space and test time, the last shift should be aligned between the wrapper scan chains and the internal core scan chains so that the complete state of the core is established at a single point in time (this may require filling or padding the beginning of short scan chains with logic Xs). The test wrapper should also support the launching of "vector pairs" into the core (and into the UDL). In this manner, deterministic scan vectors can be ATPG generated for both AC and DC fault models of the core and the core interface. These include the core's I/O specifications of "input-setup" and "output-valid."

In a similar manner to the internal scan in the core, the wrapper scan chains should be optimized for shift-bit length as well. Ideally, the wrapper scan chain bit depth should be balanced to the size of the internal core scan chains—but the same trade-off applies about the number of scan chains versus the scan chain depth—as many scan chains as can be supported (added to the core's test port) to reduce the vector depth should be supported.

5.17.3 Embedded Memory BIST

A memory BIST for every memory structure within the core should be supported (RAMs, ROMs, EEPROMs, and even some register files if they are not scanned). The memory BIST allows the embedded memory arrays to be tested algorithmically with a reduced signal interface. For example, a minimal BIST test interface would affect the core's test port with three signals that include the "BIST Invoke" input signal, and the "BIST Done" and "BIST Fail" output signals. If retention testing and characterization are to be supported by the core memory arrays, then the interface may grow to six signals with the "BIST retention" and "BIST Bitmap" input signals and the "BIST Bitmap Out" output signal.

The embedded memory BIST not only reduces the number of interface signals, but it also allows test time optimization in that all memory arrays can be tested simultaneously if the silicon or package power limits allow (it is up to the core provider to deliver the test power rating of the core and the core test vectors, but it is up to the integrator to create a final chip-level architecture that does not exceed the chip or chip package power limitations).

5.17.4 Other DFT Features

In addition to active test features such as a scannable test wrapper, an internal full-scan architecture, and memory BIST, the core should also meet the requirements of static design so that it may be a candidate for Iddq (quiescent current) testing. In general, the core should not be designed with any active logic such as pullup or pulldown logic, and a method should be provided by which the clock to the core can be stopped without loss of state.

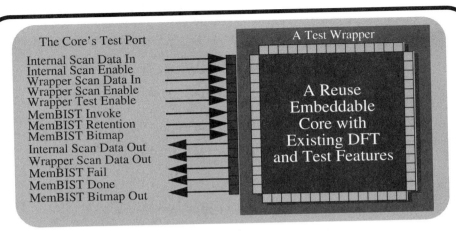

- Core Economic Considerations

Test Integration (Time-to-Market)

Core Area and Routing Impact (Silicon/Package Cost)

Core Power and Frequency Impact (Package/Pin Cost)

Core Test Program Time/Size/Complexity (Tester Cost)

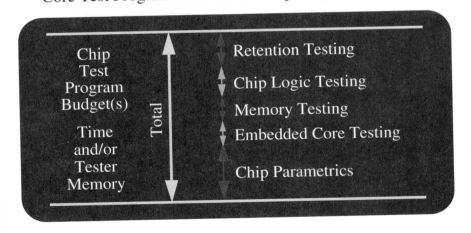

Figure 5-18   Core Test Economics

5 • Embedded Core Test Fundamentals

5.18 Core Test Economics

5.18.1 Core DFT, Vectors, and Test Economics

If a non-mergeable reuse core is designed and the business goal is to allow it to be integrated by a different fabrication house from the core provider, then the core provider will deliver a core with built-in DFT and must address the test economics of delivering a core and all of the related deliverables. The overall test economic considerations of the core design on the chip integration, as shown in Figure 5-18, include:

- the complexity of the core's test integration (TTM)
- the die impact of the core's test signal routing (TTM, Silicon Cost)
- the die impact of the core "with test" size delta (Silicon Cost)
- the package impact of the core's test signals (Package/Pin Cost)
- the package impact of the core's test power needs (Package Cost, Silicon Cost)
- the package impact of the core's test signal frequency needs (Package Cost)
- the tester impact of the core's test program size (Tester Memory)
- the tester impact of the core's test program time (Tester Socket Time)
- the tester impact of the core's test program complexity (Tester Cost)

As can be seen, the Cost-of-Test as related to a delivered core and its reuse vectors resolves into four categories: Time-to-Market impact; Silicon Cost impact; Package Cost impact; and Tester-Related Cost impact. In general, if the process is below .5 micron, and the core has been designed with a thought to the number of test signals (size of the test port), and the test feature mix is as described previously (test wrapper, full-scan, MBIST), then the dominating factor is "Tester-Related Cost." However, if no thought has been given to the core's testability, then it follows that all of the cost's are affected—but it must be noted that the tester-related cost is becoming the dominating factor in all markets.

5.18.2 Core Selection with Consideration to DFT Economics

When an organization wishes to build a chip that uses cores and the core under consideration is available from multiple sources, then the receiving organization has most likely accomplished some form of "benchmark" comparison between the various cores that are available. The comparison may be a simple data sheet comparison, or a more extensive comparison that requires establishing a set of benchmark standards. During the selection process, the receiving organization must assess the size of the core, the power consumption of the core, and even the mix and integration requirements of the test features.

One of the deliverables to the core integrator is the core vector set. In most cases, only sophisticated receiving organizations will have the foresight to request the vector set and to use the size, data rate, and overall complexity as part of the selection benchmark—and these factors may actually be the dominating cost drivers of using the core.

At the very least, early in the chip design cycle, the organization that will receive the core should establish a "test program budget" for the entire chip. This is a test time and tester memory budget. When calculated out, a test time and tester memory slot will be budgeted for the core vectors. If the delivered core vectors exceed this budget, then either some other test suffers or a greater test cost is incurred by conducting a test reload or by upgrading to a larger tester.

One of the greatest positive cost savings that can be made by the core provider is to make an effort to minimize or optimize the delivered embedded core vectors for test data sizing or test time. However, the trade-off in compressing test vectors is that they are more active in a lesser period of time, so the power consumption during application may increase.

5 • Embedded Core Test Fundamentals

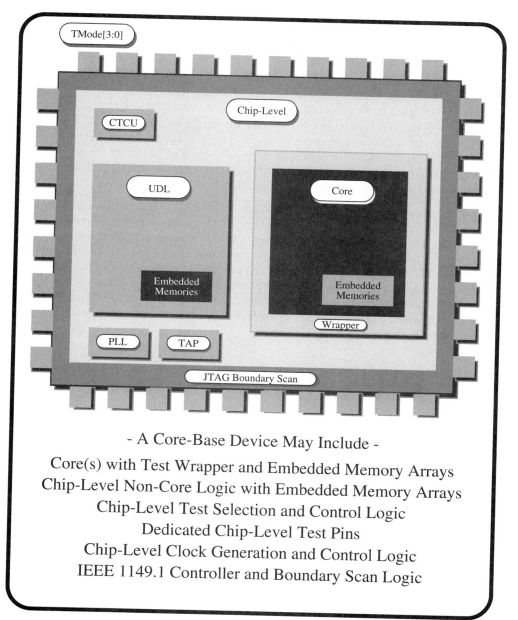

Figure 5-19   Chip with Core Test Architecture

5.19 Chip Design with a Core

5.19.1 Elements of a Core-Based Chip

It is one thing to develop a testable reuse core with DFT features; it is another thing altogether to embed the core in such a manner that the DFT features are accessible (and in the manner they were designed to be accessed), and to make the rest of the chip testable. The whole point behind core-based design is either to significantly reduce the time-to-market involved with design (actually, the time-to-volume production) or to make available enough design material to meet the customer's request for a system-on-a-chip design. The last thing the core integrator wants to do is to go through a complicated or difficult core integration, and the last thing a core provider wants is for the core to be considered difficult to integrate (or test).

Core-based design consists of integrating reuse cores, with their test features, into a chip design and adding some non-core logic. These types of chips may include non-mergeable cores with test wrappers (cores delivered with existing test logic), some mergeable core logic (cores that can have test added at the time of integration), some non-core logic (UDL), a chip-level test control unit (CTCU in Figure 5-19) to apply test scheduling between the UDL and the cores (which may each be tested in isolation or in conjunction with the UDL and each other), an IEEE 1149.1 (JTAG) architecture to support chip integration and board test, and possibly some embedded clock source (such as a PLL). These items all interact with the cores and may affect the integration of the test features.

5.19.2 Embedded Core Integration Concerns

The integration of the reuse core and the development of the chip-level access to the core's DFT should be a fairly simple and straightforward matter. If the core has been designed so that the core DFT is activated by the direct use of the test signals, then the chip-level integrator must provide the logic to supply the non-test mode default logic values on those test signals (or the core may slip into a test mode when it shouldn't). If the core has been developed with a test selection encoding and the safety defaults have been included within the core (gates to ensure that test signals remain inactive), then the integrator must only apply wire connections to the core's test signals (instead of logic) and must only provide the test mode selection signal. The chip-level test mode selection logic (if necessary to select core testing versus non-core or other core testing) may be supplied by a dedicated chip-level test controller, or the test selection may be handled by an existing standard logic such as the JTAG TAP controller and instruction register (as defined in the IEEE 1149.1 standard).

Another integration concern is the clock source. The core may need to be tested with a clock provided from the tester, which will require supporting a direct access clock signal at the edge of the chip. If the natural clock source for the core is to be an on-chip clock source such as a phase-locked-loop (PLL), then a bypass test clock may need to be supported (to provide a tester clock access around the on-chip clock source). If a bypass clock cannot be supported and the core must be operated from the on-chip clock source, then a clock-out signal taken from the core's clock tree must be provided so the tester can conduct a search to identify when to apply new data and when to observe core outputs. Note that some extra clock control logic may need to be added (such as a clock gate) to allow Iddq or memory retention testing (any testing that requires stopping the clock).

One other integration concern is the frequency of operation of the test signals on the core in reference to the chip pin interface and the data delivery rate of the tester. If the core is to be tested at a higher frequency than the pins or tester allows, then the scan shift rate might need to be applied at a slow frequency (data rate), while the sample is accomplished with a clock trick (a duty cycle change, clock muxing on the tester, or using the chip's internal clock source). If reuse vectors are delivered with the core, they may not have been generated with the limitations of "any" tester in mind—they may have been generated with a specific tester in mind. This means that the testing organization may be required to "modify" the delivered reuse vectors to meet the needs and restrictions of their testing ability.

5.19.3 Chip-Level DFT

The integrator has the task of looking at the overall chip requirements and must take all the pieces and parts described above and must put them together into an overall chip that meets the functional and test requirements. The development of a chip-level DFT architecture, in the light of delivered cores, may not be as easy as it seems. Just adding the core to the chip-level DFT adds a layer of complexity that mostly depends on the type of core being delivered. Some definitions are given below:

- **non-mergeable:** contains a test architecture that is meant to be used after integration, and therefore can't be changed.
- **mergeable:** does not contain a test architecture and can be mixed with other logic with a test architecture applied on the "whole" mixture.
- **hard core:** a layout core (that should contain a test architecture and be defined as non-mergeable, but doesn't always).
- **soft core:** a synthesizable model that can be mergeable if it contains no test architecture or can be non-mergeable if it contains a test architecture and this test architecture is meant to be used after integration.
- **firm core:** a gate mapped description that can be mergeable if it contains no test architecture or can be non-mergeable if it contains a test architecture and this test architecture is meant to be used after integration.
- **test-in-isolation:** the ability to fully test a core independent of integration (also know as the virtual test socket)—for example, a core with a test wrapper that allows the core vectors to be generated at a time before the core is submitted for integration.

If the core is a non-mergeable core, then it should have been developed to be tested in isolation (tested separately and apart from any other chip logic—the reuse vectors have been generated with the core as a stand-alone unit). If the core does allow "test-in-isolation," then the integrator can use any type of DFT on the non-core logic to meet the cost goals of the fabrication and test facility, and must think about the core only when testing the interface between the core and non-core logic.

A "test-in-isolation" core, which requires making use of the test-in-isolation features, will be tested in accordance with the restrictions placed on the core's integration by the core provider, mainly so the reuse vectors can be applied as intended. This requirement allows the chip integrator to define the testability for the rest of the chip in whatever manner makes the most sense and to install the testability for the core as it is defined by the core provider.

Two other cases may occur: the mergeable soft or firm core and the non-mergeable core that is not designed for "test-in-isolation." The mergeable core allows the core integrator to mix the core logic with the other chip logic, and the test architecture can be developed and applied at the chip-level (and the test architecture should be the most cost effective architecture based on the fabrication and test facility). This type of integration is optimal for the chip-level DFT, but the intellectual property value of the core is also given to the integrator.

If the core to be embedded is a soft, firm, or hard non-mergeable core, but has no "test-in-isolation" capability (for example, no test wrapper), then the chip integrator must create an overall chip architecture that allows both the core and the other chip logic to interact. One example would be to connect the core scan chains to the non-core logic scan chains to make chip-level scan chains that are mixtures of both. The concern with these kinds of cores is that the core integrator must have an ATPG-able design description (which gives the integrator direct access to the intellectual property content), and the core vectors can't be submitted for vector generation until the chip design is completed. The concerns here are: "What if the core has bad fault coverage with this particular integration?" and "What if the integrator's ATPG tool applies different DFT rules from the core designer's ATPG tool?" Either case would result in low fault coverage, inefficient vectors, or high test times).

As can be seen, the core integrations that allow the most DFT optimizations are the non-mergeable, hard "test-in-isolation" core, and the mergeable soft core. Since the mergeable soft core integration is not effectively different from a single chip development process, the remainder of the Core Test section will concentrate on the building of a chip by integrating with non-mergeable "test-in-isolation" cores.

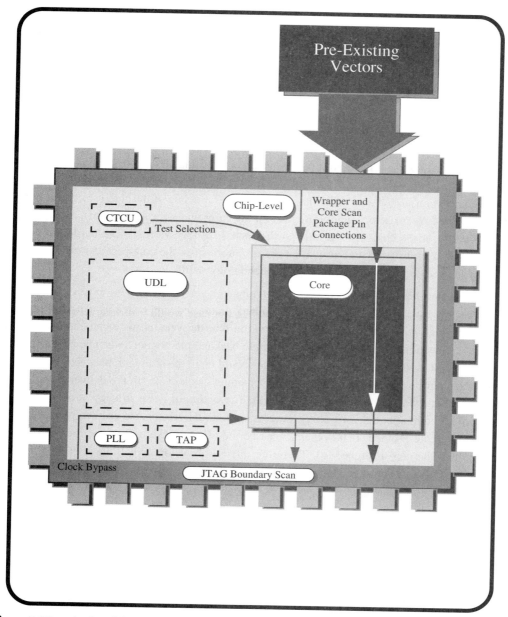

Figure 5-20   Isolated Scan-Based Core-Testing

5.20 Scan Testing the Isolated Core

The method of integration that is clever is to embed the core as a separate unit. This is generally known as non-mergeable, or "hard" core integration (because this is a must-do for a hard core), but the actual delivered core may be soft, firm, or hard. The key is that the core is embedded with a test wrapper that allows the core to be tested in isolation without requiring other chip logic to be available (see Figure 5-20). Any other method, such as integrating the core test directly to the user defined logic (UDL) test (e.g., concatenating scan chains) or developing the chip test as a unified design, is the same as a standard chip DFT effort.

The core isolation integration effort is the interesting case and is recommended for accountability purposes. In the overall business, some concerns crop up when the final product fails. The questions asked are: "Who has responsibility for debug of the chip logic?...the cores?" "Who provided which core?" (hard to separate if the cores and logic are merged), and "Who has responsibility for the chip when it fails on the tester?...in the system?" The hard core method also provides a measure of time-to-market optimization, since the core provider can deliver the core and the vectors from a core library or a "technology shelf."

When the chip is in core test mode, the wrapper is active, the core is active, and any PLL should be bypassed. The core should be integrated, connected, and tested as if it were a stand-alone item—and in exactly the same configuration that the vectors were generated for it. All scan test connections should resolve to be simple package pin connections when the part is in Core-Scan Test mode. No extra sequential logic should be in the scan path between the core and the package pins, since extra registration means adding clock cycles to the delivered reuse vectors (which will require modifying the vectors). The optimum reuse strategy is not to have to modify the vectors after delivery. Note, however, that all delivered reuse vectors, no matter how complete the core provider makes them, must be modified to take on the specifics of the chip within which the core is embedded. Therefore, the reuse vectors, which have been generated with the core's signals must be "fattened" to now represent the chip's signals, and to include any test mode selection sequences.

In one case, however, the core may be connected as intended with every intention to reuse the vectors, but the delivered vectors won't work. This occurs when the core's frequency/data-rate requirement exceeds the chip's package pin rating or the tester's frequency/data-rate capability. In this case, the decision must be made to present the vectors to the core at a slower than intended frequency (which may compromise the AC test goals) or to allow the tester to provide the scan shift data at a slower frequency/data-rate, but to allow the on-chip PLL to conduct any sample cycles or to play tester games such as multiplexing two channels onto one package pin (and altering the reuse vectors to place every other bit into a different channel). The core "test frequency" is definitely one of the early problems the chip integrator should be aware of and should try to solve in the most economically feasible method.

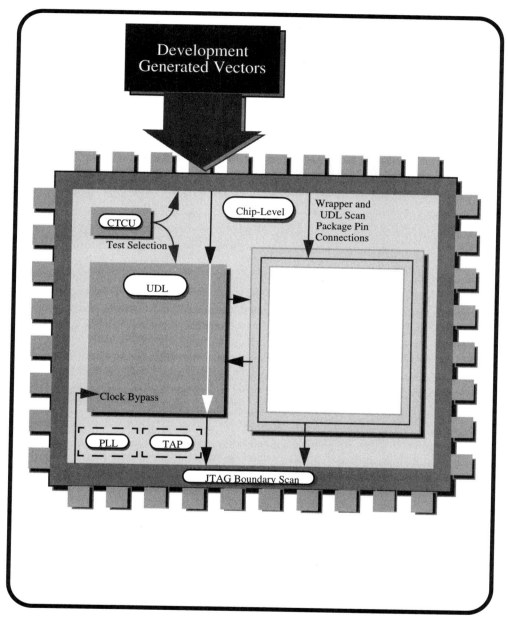

Figure 5-21   Scan Testing the Non-Core Logic

5.21 Scan Testing the Non-Core Logic

5.21.1 Scan Testing the Non-Core Logic in Isolation

Whereas the core vectors should be delivered with the core, the non-core or UDL vectors are generated after chip development (unless the UDL is pre-designed as a logic core—then this is a case of multiple core integration). To make the chip-level ATPG easier, it is not necessary to re-generate core vectors. In most cases, the customer will not even receive a gate-level design description of the core with which to conduct any sort of ATPG; the customer will just receive an encrypted HDL model for simulation purposes. However, for ATPG purposes, the core should be represented by its test wrapper. This is the shared gate-level interface (shared because the core uses it for isolated vector generation, and because the UDL uses it when the core is not present).

Note that the chip-level pin specifications (input-setup and output-valid) must be tested using the UDL part of the test vector development, not the core's reuse vectors. This is because the core vectors are generated with the core in isolation—it is not known at that time what package, pins, or integration environment will exist. This makes for one problem that is still not fully solved: "What if the core signals are connected directly to the package pins? How are those pin specifications generated?" Right now the answer is to generate functional vectors using the encrypted version of the core's functional model, or to have the core provider generate more vectors against the whole chip, including the core, after the core has been integrated.

When the UDL is being tested, the UDL needs to be active, the test wrapper needs to be active, and any PLL should be bypassed (see Figure 5-21). Note that the core should be in some quiescent default or safety state, or it should be placed in a scan mode with logic 0s being fed into all inputs and constantly shifted (the sample, or SE de-assertion, should be denied).

Note that the core and the UDL may be tested simultaneously if there are enough package pins to allow both test architectures to be used simultaneously and if power consumption is not a concern (if so, then the safe state should be applied to the core during UDL test).

5.21.2 Chip-Level Testing and Tester Edge Sets

The cost-of-test is lower if the test process requires the support of fewer edge sets (the maximum optimization occurs if all testing can be done with only one edge set). The timing point, with respect to the reference clock, at which new information is applied to a device-under-test input pin or at which the response data is observed on a device-under-test output pin establishes a timing edge set. If a package pin supports multiple application or strobe times, or if the package pin supports application or strobe times based on multiple reference clocks, then that pin is said to require more than one edge set. Another way of saying this is, "If the tester must support more than one timing standard to conduct testing on any given pin, then this is another edge set." The overall test integration should attempt to make the core connection timing and the UDL connection timing similar enough to use the same edge set.

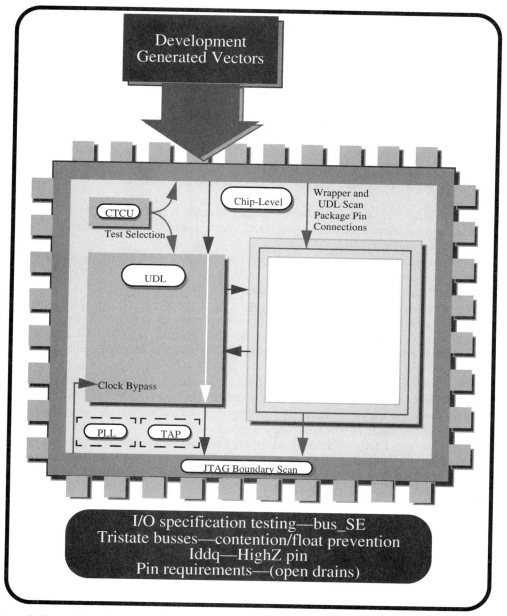

Figure 5-22 UDL Chip-Level DFT Concerns

5.22 User Defined Logic Chip-Level DFT Concerns

If the UDL is going to be tested aggressively, similarly to the core, then some chip-level techniques may need to be applied. For example, chip-level Iddq testing requires that both the UDL and the core be placed in quiescent modes simultaneously with vectors applied that exercise the pseudo-stuck-at or toggle fault models. Any active elements in both areas should be disabled. On the UDL side, which includes the chip pins, there may be a need to tristate the pins to conduct the Iddq or pin leakage tests. In any case, "quiet vector selection" is compromised, since the selected core vectors or the selected UDL vectors must be merged and each was selected and evaluated "in absentia" of the other—the overall quietness of the combined Iddq test may have to be assessed by the integrator. Sometimes extra chip features may support or make this assessment easier. For example, a single chip pin may need to be dedicated or borrowed from a functional pin during a test mode, to allow an efficient single pin HighZ or "tristate all pins" function.

Scan applied to the UDL logic must meet the same rigors as scan applied within the core, so asynchronous element control, clock control, tristate control, and other scannability rules must be met (see Figure 5-22).

If the pin specification will be tested by scan as well, then the at-speed AC scan support of a dynamic scan interface control, for example, a bus_SE (as described in the scan section 3.22.1.2), may also have to be designed. This may affect the core integration if the scan ports are used for both the UDL and the core—care must be taken not to place additional timing or pin gating requirements or restrictions on the core scan connections just because they share pins with the UDL scan ports. Differing requirements on the same pins used for UDL and core testing may result in creating the need for different edge sets to be applied when testing the separate logics.

When the overall DFT and test strategy for the chip and the UDL logic is established, any chip-level signals that interact with any scan interface that is shared by both the UDL's scan interface and the core's scan interface need to be prioritized and separated. For example, there might be a core test mode in which the scan interface pins are just "dumb" direct connections; there might also be a UDL test mode in which the same pins are now dynamic ATPG connections that allow safe shifting when SE is asserted, and pin functional sampling or strobing when SE is de-asserted; there may even be a mode in which both the core and UDL are active and being fed from the same pins simultaneously (but only the core or the UDL is driving the outputs).

5 • Embedded Core Test Fundamentals

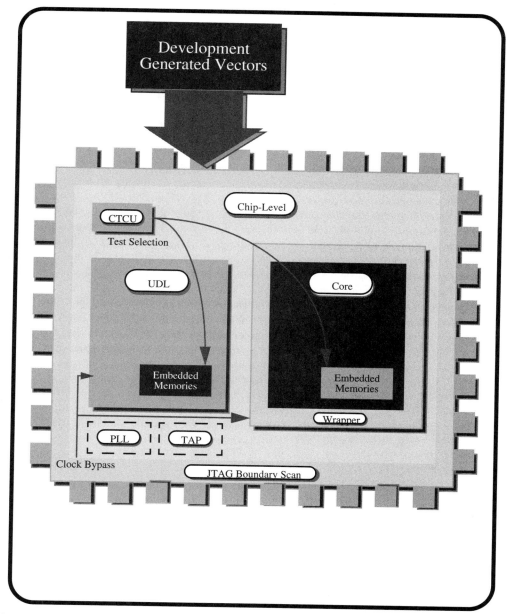

Figure 5-23   Memory Testing the Device

5.23 Memory Testing with BIST

Unless there is a power issue (usually a package issue, but sometimes a silicon issue if the power grid is insufficiently designed), all memory BISTs should be run simultaneously (BISTs in both the core and the UDL). This approach minimizes the overall test time, and allows whole chip memory retention tests to be done with only two pauses, DATA and DATABAR. If power is an issue, then the memory BISTs should be banked to stagger memory operation to within the chip's power limitations (see Section 4.24 *MBIST Power Concerns*).

Simulation to create the vectors that launch the BIST can be done at the chip level with the encrypted simulation models, so there is no need to isolate the core during this test generation (e.g., no tool other than a simulator needs to operate on the encrypted model or netlist—so the core's encrypted description can remain in the chip-level design description). The generation of the vector for the tester can also be done at the chip level, since the driving values to control the BIST and the expected response can be both applied and generated from the encrypted model.

Sometimes, however, the core provider may wish to provide the complete test suite of the core, so the memory test and logic test vectors may all be provided as reuse vectors. In this case then, the core memories and the UDL memories may be tested separately and the integration logic needs to separate these two test modes. However, a core integrator should request that the scan and memory test vectors be delivered as separate pattern sets if the test goals are to have all the memory arrays test simultaneously

Note that the core may be delivered with embedded memory arrays, and when the core is embedded, the memory arrays are "doubly embedded" (see Figure 5-23). The most cost effective method to test these memory arrays is memory BIST. However, memory BIST does not need to be applied to all memory arrays at the chip level in the UDL. If some memory arrays are connected directly to the package pin interface functionally, then a tester memory test function can be used to test these memory arrays in a cost effective manner.

5 • Embedded Core Test Fundamentals

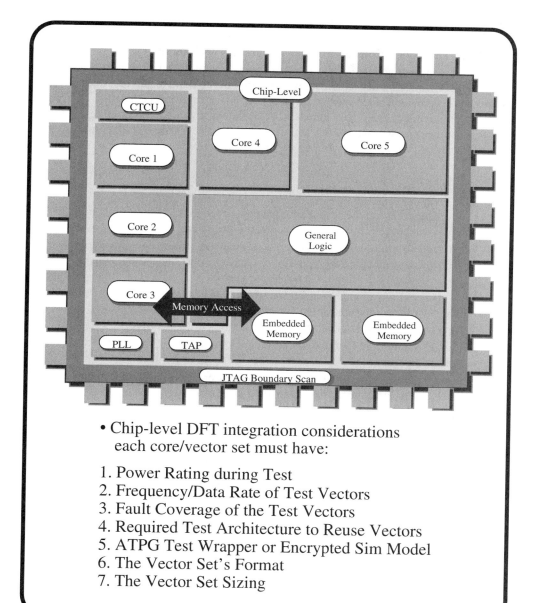

Figure 5-24   DFT Integration Architecture

5.24 Chip-Level DFT Integration Requirements

5.24.1 Embedded Core-Based DFT Integration Architecture

The chip's system architect, who must integrate any delivered core, in most cases is not from the same organization that provided the core. In order to build a chip that may include several cores and several embedded memory arrays, and to craft an overall chip-level DFT architecture, the system architect must have some key data associated with each core (see Figure 5-24). The system architect must create a DFT architecture that achieves the overall chip-quality goals and meets the final cost-of-test goals (as well as the functional engineering design budgets). Meeting these goals can be "engineered" by selecting test integration options that allow test time or test vector depth to be minimized. Some of the options the system architect must consider are:

- How many cores can be tested simultaneously to reduce test time?
- How many cores can be tested simultaneously within the final chip's power budget—what is the chip's "simultaneous" power budget?
- What kind of chip-level test mode access will be required (combinational, sequential, state machine, JTAG) to allow multiple core and overall chip-level test selection and still allow the use of reuse vectors?
- How many chip-level interconnect signals are required for each core in order to provide test access to each core?
- What are the required frequencies of the chip-level test interconnect—do cores require a frequency higher than the tester can provide or higher than the rating of the package pins?
- What is the order of testing (which cores and other chip logics are tested first, next, etc.) to provide the most efficient chip-level test flow?
- What is the test time or tester memory depth budget or requirement?

The system integrator can't begin to answer these questions unless the cores are delivered with some key test information.

- What is the maximum power rating of each core's test vector set?
- What is the maximum frequency requirement of each core's vector set?
- What is the fault coverage associated with each core's vector set?
- What is the test architecture configuration stipulated by each core provider to allow application of each core's reuse vectors?
- What ATPG model or test wrapper is supplied with each core to assist with or to allow ATPG of the non-core logic?
- What is the vector format of each of the delivered core's reuse vectors?
- What is the vector depth, width, and vector application time associated with each core's reuse vectors?
- What are the clock tree or clock skew requirements of each of the cores?

Based on this information, the system architect can then begin to create a chip-level test architecture that begins with selecting a number of package pins to dedicate or borrow (shared functional pins) to supply test data and control (access) to each of the cores. These signals must be matched to the frequency or data rate requirements of each of the core's reuse vectors. The pin interface and the power limitations applied by the silicon or the package will establish how many, and which, cores can be tested simultaneously. If conditions can't be met then the tradeoffs of adding pins or increasing the tester memory may need to be assessed. Once the pins and the simultaneousness have been determined, then a chip-level test controller (CTCU) can be developed to mediate the test selection.

5.24.2 Physical Concerns

The integration of one or several cores is more than just placing a core in some location on the floorplan and then connecting the defined test ports to the package pins. If the delivered core is a mergeable soft core, then the integrator must ensure that the core logic and the core's test logic pass through the synthesis process and meets the area and timing budgets. An HDL description may make static timing, but it must successfully pass through the physical place-and-route process to become a real and manufacturable piece of silicon. If the merged core and UDL logic are very complex, this outcome is not always possible. In some cases, even a non-mergeable soft core may pass static timing, but can't achieve physical timing convergence in the place-and-route tool.

In order to ensure that the core does meet its Max-Frequency goal, many core providers deliver hard macrocells. However, this delivery doesn't mean that the job is over. Physical concerns must be addressed even if the core is a pre-timed hard physical layout description. For example, a great concern to the test architecture is that there are no clock-data races between scan flip-flops during shifting and during sampling. This means that clock skew management must be accomplished within the UDL, within the core, and between the UDL and the core's test wrapper. The skew management is independent of the frequency of operation; it is wholly dependent on the clock-to-out speed of the individual flip-flops versus the skew inherent in the clock trees.

What makes clock skew management critical with cores is that a hard core has a pre-made clock tree of its own. This clock tree must be merged with the overall clock structure of the chip. If the core has the same chip-level clock source as the UDL, then the core's clock tree must be directly merged with the chip clock tree. If the core has four levels of inversion in its clock distribution network, and the UDL will have six, then the thought process is to connect the core to the second level of the chip's clock tree. However, the core's clock distribution network may not be made the same way—the core's clock tree may be more or less efficient than the overall chip's clock tree, leading to a delay difference that may be a skew problem.

The other physical concerns are the ability to meet the area and timing budget with the additional chip-level routing of the core test ports to the package pin interface for test access and the sizing of the power grid to support the test (for example, the power grid must supply enough current to the core to support the power requirements of at-speed shifting.)

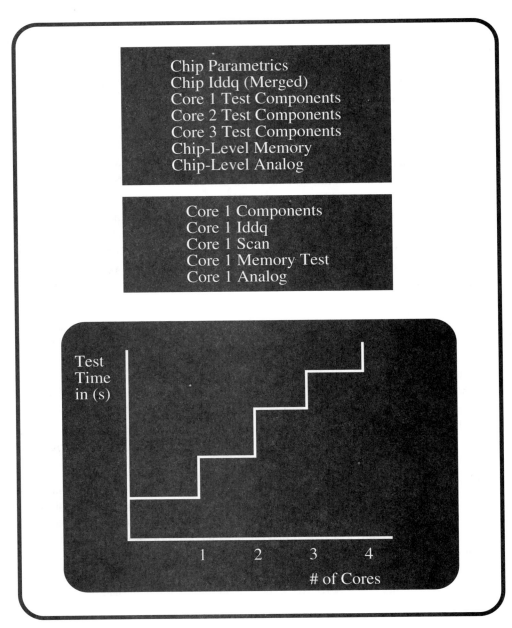

Figure 5-25 Test Program Components

5.25 Embedded Test Programs

The test program in current IC testing generally contains more than just the "functional verification vectors" that are delivered by the design team. Most test programs contain the same type of vector products or test patterns groupings. Note that the purpose of a manufacturing test program should be for structural verification of the silicon and not for "functional design verification." The test program should be "economically optimized" to provide a cost effective and efficient method to verify that the manufacturing process did not add errors or defect content to the finalized design. The test program should not be used as a vehicle to verify the final design or design library assumptions (unless the chip under test is the process technology vehicle).

Most chip designs contain the following general vector products, which are usually arranged in an order to provide an early exit if the test program finds a failure—this is known as stop-on-first-fail and can significantly increase tester throughput if the test program is long compared to the tester's handler index time (see also Figure 5-25).

- Chip-Level Pin Parametrics
- Test Logic Verification
- DC Logic Stuck-At
- DC Logic Retention and Iddq
- AC Logic Delay
- AC Frequency Assessment
- AC Pin Specification
- Memory Testing
- Memory Retention

The difference between a unified chip design test program and a core-based design test program is that a portion of the test program may come from the core provider. If the core is proprietary, then the integrator may receive only an encrypted simulation model that cannot be used effectively for structural vector generation. The result is that the test program is now assembled from core vector components from different providers (the 40 core problem—merging 40 vector sets from 40 different sources). On the one hand, this may make test program assembly easier if the integrator demands a known industry standard vector format such as WGL, VCD, or STIL. However, on the other hand, the integrator no longer has direct control over the resultant test time or the length of the test program (amount of vector data that must fit in the tester's memory) unless the delivered vectors are competitively assessed as part of the business.

For example, if two cores are very similar in size and function and their pros and cons are weighed for selection, then the size of the vector set for a given coverage level should also be assessed as part of the selection process. A $3.00 core with a $5.00 vector set is not as good a deal as a $7.00 core with a $.50 vector set. This situation can be dealt with if the integrator designs the overall chip test environment to make use of vector parallelism (schedules test components to run simultaneously).

- Receiving Core DFT Specification

- Driven by Fab and Integration Requirements

- Core DFT Specification Items
    - Test Mix
    - Style of Test
    - Maximum Number of Integration Signals
    - Minimum-Maximum Test Frequency
    - Maximum Vector Sizing
    - Minimum Fault Coverage
    - Clock Source

Figure 5-26   Selecting or Receiving a Core

5.26 Selecting or Receiving a Core

When selecting a core to include within a design, the integrator or receiving organization, needs to establish the bounds of acceptance. At the very least, there should exist a receiving specification, developed by the core integrator, that defines all of the requested test modes and test integration requirements. For example, if the integrator plans on developing a chip with multiplexed-D flip-flop scan, then an incoming core with modeled LSSD scan is not compatible and would be difficult to integrate into the test architecture.

The specification should be driven by both the needs of the integrator (the design and build environment) and the needs of the fabrication facility (what is the required or most economical test application for the fabrication facility and test floor). For example, the test floor may have only testers that operate at 50MHz with a maximum of 2 MB of memory depth on most channels, but with 16 channels of 16 MB "scan option" memory depth—this would drive the core DFT solution to be 16 or fewer scan chains and a memory BIST methodology, since the functional vector space is limited.

The incoming core DFT specification, summarized in Figure 5-26, should define:

- the type of core (mergeable, non-mergeable, hard, soft, firm)
- the type of test required (scan, BIST, functional, pseudo-functional)
- the style of test (MuxD Flip-Flop, LSSD)
- the maximum number of test integration signals
- the required fault coverage
- the types of fault coverage (stuck-at, transition delay, path delay)
- the specialty test support (Iddq, Idd)
- the budget/limit on the number of vectors or test time
- the maximum frequency/power of test operation
- the test clocking requirements (bypass, PLL-only)
- the core documentation standard (power, vector format, test operation, etc.)

The specification, once developed, should be one element used as the basis for core selection or core benchmarking (other functional and physical core criteria such as size, functional power consumption, clock frequency, etc., should be documented and used as well). The receiving organization, when faced with multiple choices for a "core socket," should select the core that most matches the specification or represents the most economic set of choices.

- Core Test Driven by Cost-of-Test and TTM

- Two Concerns: Reuse and Integration

- Reuse: Interface, Clocks, Test Features
    - number of dedicated test signal
    - size of test integration interface
    - ability to test interface timing
    - no functional bidirectional ports
    - specifications and vectors based on clock-in
    - specifications and vectors based on clock-out
    - ability to stop clock for retention or Iddq
    - number of clock domains
    - at-speed full scan
    - at-speed memory BIST
    - use of a scan test wrapper
    - self-defaulting safety logic

- Integration: Core Connections, Chip Test Modes
    - simple core integration
    - reuse of pre-existing vectors
    - application of test signal defaults
    - shared resources (pins and control logic)
    - shared testing (parallel scheduling)
    - chip level test controller

Figure 5-27 Embedded Core DFT Summary

5.27 Embedded Core DFT Summary

This chapter has pointed out many, but not all, of the test and DFT concerns of making a reusable core product, and of designing with the core-based design methodology (see Figure 5-27). The key items taught in this chapter were:

Cost-of-Test and Time-to-Market are the big drivers of embedded core design and core-based design. The eventual goal for a core provider is to supply easy-to-integrate cores to the system-on-a-chip or rapid design markets. Since the vectors may now be delivered with the core, the size of the vector file is a competitive pricing item.

The two major concerns in the core-based design methodology are a reusable core and the integration of a core into an overall chip. If the core is small and not of a great complexity, then it can be integrated by bringing all the functional signals to the edge of the package within which it will be embedded. However, for a core of any complexity, it is better to include some test features to simplify integration and to reduce the overall cost of test.

The main elements that must be considered when creating a reusable core are the integration test interface, the clock requirements, and the inherent DFT features. For the core, as a competitive business, these elements must be minimized or optimized so that they will be easily usable by the general market. This means that the number of signals to be brought from the core to the chip package pins should be minimized. However, from this minimum interface, a very high-quality measurement of AC and DC fault models should be possible with some minimized test time or test data sizing, since the vectors may now be part of a larger test program. This requires attention to the source clock for test—it is best to provide it directly from the tester, since this will allow specification measurement from a clock-in, and it will simplify clock stopping for retention and Iddq testing.

The main DFT concerns during the integration of a reusable core are that the DFT features of the core be integrated in such a way as to operate as intended—and with "simple" integration techniques—and that the core test features be integrated into the overall chip level test so as to minimize the cost of test. This concern generally means that the chip must support a test control unit to select and schedule the various core and chip test features, and that some thought must be given to the sharing of resources and sharing of test scheduling to reduce the physical impact and to reduce the overall test time.

Ultimately, the huge designs of today will most likely turn into tomorrow's standard cells. A core developed today must easily be mapped to new processes and must be easily integrated by others with minimum "hand holding." When a core is delivered, it should be delivered with a "data sheet," an "integration guide," and a set of "pre-existing vectors" that are designed to verify what is on the data sheet. The integration guide should explain both the functional and the "test" integration of the core.

5.28 Recommended Reading

To learn more about the topics in this chapter, the following reading is recommended:

Abromovici, Miron, Melvin A. Breuer, Arthur D. Friedman. *Digital Systems Testing and Testable Design.* New York: Computer Science Press, 1990.

Digest of Papers, 1st IEEE International Workshop on Testing Embedded Core-based Systems. Los Alamitos, CA: IEEE Computer Society Press, 1997.

Digest of Papers, 2nd IEEE International Workshop on Testing Embedded Core-based Systems, Los Alamitos, CA: IEEE Computer Society Press, 1998.

IEEE Standard Tests Access Port and Boundary-Scan Architecture. IEEE Standard 1149.1 1990. New York: IEEE Standards Board, 1990.

Parker, Kenneth P. *The Boundary-Scan Handbook, 2nd Edition, Analog and Digital.* Norwell, MA: Kluwer Academic Publishers, 1998

Rajski, Janusz, and Jerzy Tyszer. *Arithmetic Built-In Self-Test For Embedded Systems.* Upper Saddle River, NJ: Prentice Hall, 1998.

Roy, Kaushik, ed. "A D&T Roundtable on Testing Embedded Cores," *IEEE Design & Test of Computers,* Vol. 14, No. 2: April-June 1997, pp 81-89.

van de Goor, A. J. *Testing Semiconductor Memories: Theory and Practice.* Chichester, U.K.: John Wiley and Sons, 1990.

About the CD

The Materials

The CD included with this book contains full color versions of all of the graphic figures in the book. In their original form, the graphic figures were the slides of the Design-for-Test class that I have been teaching for the past few years to designers, managers, customers, budding test and DFT engineers, and university students. These slides are provided for use as colored versions of the text graphics to assist in understanding the material (some of the highlighted points are much clearer when color is used to stress the point). They can also function as instant teaching or support materials for classes, presentations, discussions, and the task of applying influence for DFT purposes (it is always good to have a slide or two handy when you have to justify the impact of scan, BIST, test pins, etc., on chip area, frequency, power, package, and test budgets).

The use of these slides for personal study, design presentations, management presentations, engineering discussions, and internal training presentations is encouraged; however, these slides can't be used as the basis for a DFT for-profit class without the express written permission of the publisher.

The material on the CD is arranged by the five chapters of the book, and in the same order that they appear in the book (the figure numbers are identical). The slides associated with the five chapters of the book can be categorized as follows:

1. Test and DFT
2. ATPG
3. Scan
4. Memory Test
5. Embedded Cores.

In Chapter 1, there are eleven slides (1-1 to 1-11) that present the material of: defining test, DFT, measurement criteria, fault models, types of testing, test equipment, test program components, pin timing, and other fundamental concepts of test and DFT—including the "cost of test," concurrent test engineering, and the trade-offs and pros and cons of DFT. These slides can be used to present the basic concepts of testing, what is being tested, and why.

In Chapter 2, there are eighteen slides (2-1 to 2-18) that present the material involving: the adoption and explanation of an ATPG process, the tasks used in accomplishing ATPG, the software analyses applied to conduct ATPG, and the benchmarking process used to select the ATPG process and tool that is right for the business case. These slides can be used to present the basic concepts of ATPG, how ATPG can be implemented, and the trade-offs that may make an ATPG process a more cost effective method than other historical vector development methods.

In Chapter 3, there are twenty-eight slides (3-1 to 3-28) that present the material involving scan and logic BIST. This material includes: the basic fundamentals of scan, the rules involved with the adoption and application of scan, the techniques applied to overcome design limitations (scan rule violations), and some intermediate and advanced topics such as at-speed scan, using scan for AC test goals, dealing with clock-skew and power issues exacerbated by test, selecting critical paths for delay-fault based testing, and using automatic test pattern generation and compression in conjunction with a scan architecture.

In Chapter 4, there are twenty-eight slides (4-1 to 4-28) that present the material involving memory test, memory interaction with scan, and memory BIST. This material includes: the basic fundamentals of memory testing, detection and isolation of memory defects and faults, algorithmic test generation, bit-mapping for diagnosis, scan black-boxing methods, and memory BIST development and integration techniques.

In Chapter 5, there are twenty-seven slides (5-1 to 5-27) that present the material involving developing a testable reusable core and integrating a testable reusable core. This material includes: the definition of a reusable core, core-based design, and "virtual test socket" test in isolation; and also the trade-offs of core-based design and the different integration methods and techniques that can be applied.

This material is all "true" presentation material in that the diagrams do not tell the whole story (such slides would be too complicated and busy to be easily viewed and understood). These slides are meant to be used in conjunction with the material in the book to flesh out the meaning.

Technical Information

The CD-ROM is formatted for use with UNIX® and Microsoft® Windows platforms. The materials are presented in Adobe® Acrobat® PDF® format, which is platform independent.

Adobe Acrobat Reader software is included on the CD. Installers are offered for the following platforms:

HP-UX®:	`hpux/hpux/ar302hp.tar.Z`
Solaris™:	`solaris/solaris/ar302sol.tar.Z`
Sun OS™:	`sunos/sunos/ar302sun.tar.Z`
Windows NT®:	`nt/ar40eng.exe`

If your platform of choice is not represented, you may download the appropriate Acrobat Reader software for free from `http://www.adobe.com`.

Acrobat Reader allows you to view the materials interactively on screen, starting with the file Home.pdf. This is a live table of contents with clickable links to the other files. The individual chapter files may also be accessed directly, using the naming convention ch01art.pdf.

For teaching, you may display the slides directly from the files, using a projection screen linked to your computer, or you may print the PDF files to acetate sheets for use with a standard overhead projector.

Support

Prentice Hall does not offer technical support for any of the software on the CD-ROM. However, if the CD itself is damaged, you may obtain a replacement copy. Send an email describing your problem to:

disc_exchange@prenhall.com

Glossary of Terms

AC Fault Models: Mathematical descriptions of the behavior involved with assessing timing compliance of a circuit—the most common examples are the gate-level transition delay and the path delay fault models that are based on slow-to-rise and slow-to-fall logic (voltage) transitions.

AC Parametric Testing: A type of testing based on measurements of the timing sensitivities of the input and output pins (the input-setup, input-hold, output-valid, tristate-valid, and output-hold).

AC Pin Specifications: see *Pin Specifications*.

AC Scan: A subset of at-speed scan where a scan architecture supports the ability to conduct an at-speed sample cycle for the purpose of verifying timing compliance.

AC Scan Insertion: The installation of a scan architecture into a design description by substituting scannable flip-flops for functional flip-flops, and by installing the scan data signals, SDI and SDO, and the scan enable signal, SE, and any other necessary scan control signal, in such a manner as to allow the final scan architecture to assess timing compliance—this type of scan architecture must support the ability to conduct a sample at the rated frequency and generally supports the ability to shift at speed.

AC Testing: The application of vectors to a circuit with the intent to verify the timing compliance—for example, testing using the transition or path delay fault models to determine the delay or cycle-time involved in conducting register-to-register, input-to-register, register-to-output, or input-to-output data transfers.

ASIC: *Application Specific Integrated Circuit*—a specialized integrated circuit designed and optimized for a specific purpose or function as opposed to a generalized or standard product integrated circuit such as a general purpose microprocessor.

ATE: *Automatic Test Equipment*—a class of equipment used to apply test vectors to a completed chip design—also known as a tester, test platform, or IC ATE. A tester may be used to test a final packaged product, or it may be used in conjunction with other support equipment such as wafer probers and E-beam probers to conduct test on raw die.

ATPG: *Automatic Test Pattern Generation*—the use of a software tool to generate test vectors algorithmically.

Abort Limit: A limit applied to pattern generation to place a restriction on the amount of time or CPU activity involved with creating a test vector or set of test vectors.

Accessibility: Having the ability to apply vectors to control points or to observe response data at observe points—this is especially critical for embedded devices where test features may be buried within a chip design and can't be used unless an access method is provided.

Access Time: The time period taken by a memory array to present valid output data after a read has been requested.

Address: A location in a memory usually related to a selectable "word." *Address* may also mean the binary value that represents a location in a memory array.

Address Bus: The wire routes that deliver the logical address to a memory for the purpose of selecting a location within that memory.

Address Decode: The logical resolution of a binary value on an address bus to a selected "row" or "word" within a memory structure.

Address Space: The set of addresses that validly select locations within a memory, or the collection of addresses that have valid memory, or other addressable devices, connected and selectable.

Aliasing: The condition that two or more errors will self-repair and result in the "good" signature in a pattern compression device such as a MISR LFSR—aliasing is generally represented as a probability related to the bit-length of the pattern compressor (2^{-n}, where n is the number of bits).

Algorithm: A recursive computational procedure or repeating process that is applied through software to many design and test problems such as test pattern generation, scan insertion, or scan routing optimization.

Aspect Ratio: The width-to-height ratio of X-length to Y-length of layout elements on a floorplan—for example, a hard core megacell with a 1:1 aspect ratio would be a square element, and a memory array with a non-1:1 aspect ratio would be a rectangle.

At-Speed Scan: A scan test architecture that allows the scan chains to be operated (shift and sample) at the rated clock frequency for purposes of verifying timing criteria (frequency, specifications, AC faults) and of reducing the tester application time. Even though this definition is more correctly applied to allowing shifting and sampling at the rated frequency, sometimes this definition is used interchangeably with AC Scan, which is required only to allow samples at the rated frequency.

Backtrace: The tracing of nets, signals, gate-inputs, gate-outputs, and other design element component connections backwards through a design description (from outputs toward inputs). Sometimes referred to as *Backtrack*.

Backtracing: The act of tracing backward through a design description with decisions and recovery made for thwarted tracing pathways. Sometimes referred to as *Backtracking*.

Backtracing Limit: A limit applied to the number of backtracing recovery events during vector generation to minimize the amount of time or CPU activity involved with a difficult to generate vector. Sometimes referred to as *Backtracking Limit*.

Banked Testing: A testing technique where units under test are not all operated simultaneously, but are arranged in "banks" to minimize the peak power consumed during testing. For example, 15 memory arrays, each with its own BIST, may operate in 3 banks of 5 memory arrays, where the banks are sequenced one after the other, to minimize the peak power consumed during the testing of all 15 memory arrays.

Benchmarking: The comparison of items against a "best in class" standard or against each other with a "selection criteria" established. This term can be applied to any tools, processes, or devices. For example, in the case of DFT, it is usually applied to the process of comparing various ATPG tools, with an established selection criteria, to choose the tool that best meets the needs of the design or test organization—in core selection, it can be applied to the process of comparing embedded cores, with an established selection criteria, to choose the core that best meets the needs of the chip design organization.

Bidirectional Signals: Signal routes that can be used to transmit data as input or output—there are two aspects of this type of signal, "tristate nets" where multiple drivers may place data on a single wire net, or input/output nets at an interface boundary.

BIST: *Built-in Self-Test*—a design unit comprised of on-chip resources to apply stimulus and evaluate responses so a device can be tested to some level from within the chip (for example, memory BIST allows a memory test generation and verification unit to be placed on-chip, for the purpose of testing the on-chip embedded memory arrays, with minimum direct interaction from an external tester). See also *Logic BIST* and *Memory BIST*.

BISTed Memory: A memory array with BIST logic closely associated—more specifically, one with the BIST logic in such close association that the physical unit may be treated as a single design entity, with associated signals such as Data, Address, Read-Write, Output-Enable, BIST-Invoke, BIST-Done, and BIST-Fail (as opposed to, for example, a single centralized BIST controller that has the ability to test multiple memory arrays).

BIST Test Wrapper: A design unit or level of hierarchy, viewed as the external interface of a non-mergeable or "test-in-isolation" core, that uses embedded built-in self-test techniques, such as LFSRs for pattern generation (PRPG) and/or response evaluation (Signature Analysis), as the method to apply and/or evaluate embedded and isolated testing.

Bit Cell: A single memory cell that contains a single bit of storage data.

Black-Boxing: An isolation technique that allows a design unit to be removed from consideration during test vector generation. For example, a memory array or an embedded core can be black-boxed by surrounding it with a scan test wrapper—for vector generation, only the wrapper need be evident.

Boundary Scan: An access method where design units are surrounded by a scan architecture so that access to inputs and outputs may be through the scan architecture. The most common boundary scan architecture is the one based on the IEEE 1149.1 Boundary Scan standard (also referred to as JTAG).

Bridging Fault Model: A fault model represented by defining the actions that occur with the shorting together of two elements—these elements can be transistor components and connections, gate components and connections, or wire interconnect signals; and the bridge can be zero ohm, resistive, or diodic. The case of an element being constantly shorted to power (VDD) or ground (VSS) is the stuck-at fault model.

Built-In DFT: Cores delivered with pre-existing DFT—cores delivered as hard layout macrocells or megacells must have DFT installed before becoming layout blocks; similarly, non-mergeable cores meant to be used "as delivered" with reuse vectors also must have DFT installed before deployment.

Burn-In: A process by which a newly manufactured silicon product is exposed to accelerated initial operation to exacerbate infant mortality. Burn-in is generally the operation of several silicon devices simultaneously for an extended period of time while being subjected to changing environmental conditions (for example, the devices being contained within an oven). Burn-in may be applied to loose die, die on wafers, and packaged parts.

CAD: *Computer-Aided Design*—the use of computer-based tools to reduce the time or complexity of the design process, or to automate highly redundant and repetitive tasks.

CAE: *Computer-Aided Engineering*—the use of computer-based tools to reduce the time or complexity of the engineering process, or to automate highly redundant and repetitive tasks.

Circuit Learning: A process conducted by the ATPG tool to identify complex circuits or structures that are difficult to operate or make tracing decisions through, and to establish the way to deal with these circuits prior to beginning algorithmic test generation. Circuit learning reduces the time involved with the vector generation or enhances the ability of the ATPG engine to successfully create a vector.

Clock Domain: All sequential elements connected to the same physical clock distribution network—note that a chip design may have multiple domains and some domains may be defined to be operated at the same frequency; however, if the physical clock nets are distributed separately with different skew and delay management, then they are considered different clock domains.

Clock Skew: The delay difference in the delivery of a clock edge to elements at different locations on the clock distribution network—Max Skew is the maximum difference evident at some two elements in the whole clock tree. Clock skew contributes to the problem of "data smearing" or "data dropping" when a data transfer between two clocked state holding elements is faster than the clock skew between the two elements.

Clock Source: The source of the clocking signals. For example, an external package pin would enable a synchronizing "external" clock to be delivered from the tester or an external crystal oscillator, or an internal PLL or clock division/multiplication logic would result in an "internal" or "embedded" clock source.

Column Decode: The resolution of data and the read or write control signals to the individual bit cells in a memory array structure. Column decode is directly related to decoding the data locations within a word, whereas row decode is related to decoding the address and selecting the word.

Combinational ATPG: A type of pattern generation that is designed to operate most efficiently on a circuit description that is comprised only of combinational gates—there should be no state holding logic, sequential logic, or memory logic in the circuit description.

Compaction: see *Vector Compaction* and *Response Compaction*.

Comparator: A logic structure used to verify that two data values are equivalent. Comparators are used in memory BIST to verify that the data read from a memory location matches what was written into that location earlier.

Concurrent Test Engineering: The methodology whereby the test issues, concerns, goals, and budgets of a chip design are dealt with concurrently to the functional design. The opposing methodology has been called "over the wall" in that the design is accomplished and then it is "tossed over the wall" to a test group to then eke out the quality measurement on the completed and frozen design.

Cone of Logic: A descriptive term for the sum total of the combinational logic that resolves to a single signal fed to an observe point—it is referred to as "cone" because there is much more logic at the source of the signal's decode and it reduces to a single signal at the observe point. For example, when conducting scan diagnostics, a single erroneous value captured within a single scan flip-flop must have its source in the "cone of logic" feeding that scan flip.

Constraint: The limiting or forcing of certain logic values at control points. It is sometimes necessary to set certain control points (input pins, control test points, or scan registers) to particular values to enable test modes; limit, reduce, or eliminate contention; disable dynamic logic; or apply safe shift conditions—these control points then become "constrained" nodes.

Contention: A condition where two or more drivers are attempting to place different logic values on a single wire or net—this is a destructive condition and should be avoided.

Controllability: The ability to set nets, nodes, gate inputs or outputs, or sequential elements to a known logic state. This has two main contexts, the ability for an ATPG tool to place known values wherever necessary in the design description during the vector generation activity and the ability for a tester to place known values at the edge of a chip, or within a chip by using a test architecture such as scan.

Core: see *Embedded Core*.

Core-Based Design: A design methodology based on using non-mergeable cores to reduce the design cycle time, ease integration, or enable a "system-on-a-chip" or ULSI design. The key to core-based design is to accept the logic and/or memory megacells as completed design units, and to integrate them in a plug-and-play manner.

Cost-of-Test: The measurement of the expense generated during a chip design or of the cost added to a final chip product, due to the DFT or test process. The cost-of-test is described as the additional time added to the design cycle due to DFT and test activities (TTM impact); the additional time added to the product cycle due to vector generation for qualification time—time-to-volume (TTV impact); the direct impact to the silicon size due to test logic (Silicon Cost); the direct impact to the number of package pins and to the power rating of a chip package due to test logic (Packaging Cost); and the direct costs involved with the testing process, such as the cost of the tester, the throughput time on a test floor, the cost and complexity of the test program, etc. (Test Cost). These costs can be separated into recurring and non-recurring costs, where recurring costs are incurred each time a final chip is tested.

Critical Path: Those paths that establish the product's frequency, or pin specification limits, within some defined timing guardband in a fault-free circuit, and those paths that would establish the product's frequency or pin specification limits in the presence of delay faults. Critical paths are most accurately exercised and detected by vectors generated against the path delay model, and path delay vectors should be generated for all paths with timing slack that would not be caught by some other test such as stuck-at or transition delay-based testing.

Current-Based Fault Model: A fault model represented by defining the actions involved with current leakage conditions. This fault model is usually applied by supporting any of the structural, bridging, or delay fault models, to exercise the fault conditions, but the fault detection is the measuring of the current drawn by the circuit as the observe operation, rather than the propagation of voltage values to observe points.

DC Fault Models: Mathematical descriptions of faulty behavior designed to assess structural compliance of a circuit independent of any timing requirements—the most common example is the gate-level single stuck-at fault model.

DC Pin Parametrics: A test based on measurements of the voltage and current sensitivities of the input and output pins (the voltage and current levels associated with a logic 1 and 0, and the high impedance or mid-level states).

DC Scan: A type of scan architecture or operation where the intent is to apply "slow" (well below the rated frequency) scan shift and sample operations to assess the structural compliance of the design and not any timing compliance criteria.

DC Scan Insertion: The installation of a scan architecture into a design description by substituting scannable flip-flops for functional flip-flops, and by installing the scan data signals, SDI and SDO, and the scan enable signal, SE, and any other necessary scan control signal, in such a manner as to allow the final scan architecture to assess structural compliance only—this type of scan architecture needs only to support the ability to conduct scan shifting or sampling at a rate much slower than the rated frequency.

DC Testing: The application of vectors to a circuit with the intent to verify the structural compliance—for example, testing using the single stuck-at fault model.

Defect-Based Testing: A type of testing where the nature of the test is meant to directly exercise, detect, and isolate defects and defect effects rather than abstract fault models. This type of testing is applied to memories because they are regular structures and the failure modes and effects of common memory defects are well understood.

Defects: The "real world" physical anomalies that can occur in silicon or in packaging. Defects may result in the final chip product not being in compliance with its operation parameters (logic functions, frequency performance, power consumption) or may result in a limited product lifetime (reliability). Defects may exist, but may not be destructive—not all defects map to fault models and ultimately chip failure modes.

Delay-Based Fault Model: Mathematical representations of behavior based on the concept of adding delay to a circuit's timing. These models are applied by the launching of a voltage-based logic transition (0->1, 1->0) into a circuit, assuming that a slow-to-rise or slow-to-fall delay is evident, and then observing the circuit's logic response to that transition at a later time interval. The most common examples of delay-based fault models are the gate-level transition and path delay models.

Delay Fault Models: see *Path Delay* and *Transition Delay.*

Delay Path Sensitization: The establishing of a propagation pathway from one point in a circuit to another point in a circuit, through a described pathway of gate elements, by controlling the "off path" values of the gate elements in the path—for example, the AND type gate would have a logic 1 placed on the off-path gate inputs to allow the selected path-input to pass the logic 1 or logic 0 value. This analysis is done by an ATPG tool during the generation of path delay vectors.

Design Rules for Test: Also known as DFT rules—the set of design restrictions placed on a design description for the test process. These rules are applied to ease the vector generation process (manual or ATPG); to enhance the quality measurement metric (fault coverage); or to reduce the cost-of-test (test time, vector depth). Common design rule examples are: no asynchronous sequential elements, no combinational feedback paths, no non-scannable sequential elements.

Design-for-ATPG: A subset of DFT that can be described as the action of applying design rules to a chip design to ensure that an ATPG tool can successfully generate vectors to achieve a targeted quality level.

Deterministic Vector Generation: A type of ATPG where all vectors that are created are crafted deterministically by targeting faults and algorithmically generating a unique vector to exercise the fault and to propagate the fault effect to an observe point. The other form of vector generation is "random," where random groupings of logic 1s and 0s are applied and fault coverage is assessed by fault simulation.

DFT: *Design-for-Test*—the action of placing features in a chip design during the design process to enhance the ability to generate vectors, achieve a measured quality level, or reduce the cost-of-test.

Design Verification: The evaluation of the behavior, functions, operations, or other design engineering budgets of a device by the application of behavioral, functional, or operational vectors graded against a "golden standard" of behavior compliance as opposed to a "structural" metric. Design verification can be assessed in a behavioral, register transfer, or gate-level simulation, or on manufactured silicon.

DIE: see *Silicon Die*.

DRAM: *Dynamic Random Access Memory*—a memory that uses a capacitor as a storage element and requires the application of power and a "refresh" event to retain its data.

EDIF: *Electronic Design Interchange Format*—a text language or format used to describe electronic design information. See also *Verilog*.

Edge Set: see *Tester Edge Sets*.

EEPROM: *Electrically-Erasable Programmable Read-Only Memory*—a type of programmable non-volatile memory.

Effective Test Interface: The test interface of an embedded core when a test wrapper or other interface reduction device is used to optimize the access pathway to the core.

Embedded Core: A reusable design unit, meant to be integrated as part of a larger design, that can be delivered as a soft core (HDL and RTL), firm core (gate-level netlist), or hard core (layout description). The core can be further grouped as mergeable (the core can be mixed and merged or flattened with other logic) and non-mergeable (the core is meant to be delivered and integrated as a complete unit and "tested-in-isolation").

Embedded Microprocessor Access: A test methodology where an embedded microprocessor is used as the engine to conduct on-chip testing.

ESD: *Electro-Static Discharge*—a term describing the immunity a manufactured silicon part has to destructive effects due to static electricity discharges.

Exhaustive Testing: A test method whereby all possible 2^n logic values are applied to the inputs of a combinational circuit to exercise every possible applied value. For sequential circuits, this method requires the application of all possible 2^n values to the inputs, and the application of all possible 2^m sequential sequences of the 2^n values (2^{n+m} testing).

Expected Response: The predicted or deterministic output of a circuit under test. When a circuit in a known state is exercised by a known input stimulus, the cared outputs (real response) must match the expected response—any difference between the real response and the expected response is a failure.

Failure: A real vector response that is measured at the tester and compared to the predicted "expected response" and does not match. A failure is the criterion used to discard silicon product. Vectors are applied to exercise faults, so a successfully exercised fault that is propagated to a tester observe point usually is the source of a failure.

Failure Analysis: The action of relating device failures to their root cause (defects, mask errors, manufacturing errors, design errors, etc.) and of developing corrective actions to "enhance yield," or raise part quality.

Fake Word Technique: A method of bypassing a memory during scan testing by providing a register or single memory location to emulate the presence of the memory—this allows vector generation to be accomplished without including the memory in the design description acted on by the ATPG tool.

False Paths: Path tracings within a design that can't be exercised by functional vectors. False paths are a problem when attempting to generate path delay vectors for AC verification—there can be many more paths than can be economically tested if the ATPG engine can generate vectors for "false functional paths;" there might be difficulty in identifying the true paths that limit the frequency of the part; and the "effective delay fault coverage" may seem low if the ATPG engine can't generate vectors for false paths.

Fault Coverage: The metric of how many faults are exercised and successfully detected (their fault effect is observed) versus the total number of faults in the circuit under test.

Fault Dropping: The part of the ATPG process where the tool removes detected faults from the total list of faults contained in the circuit. This is a tool runtime and a vector sizing optimization operation.

Fault Detection: The observation of a propagated fault effect at a legal "detect" point. A legal detect point is one that can really be observed by a tester, not one that can be seen only in a simulator.

Fault Effective Circuit: A representation of the circuit where the logic is re-mapped based on the fault effect. For example, a stuck-at one on the input of a logic gate may eliminate that gate and other downstream logic. This type of circuit is a result of a failure modes and effects analysis.

Fault Enumeration: A step in the ATPG process where the faults associated with a design description are identified and listed so that they can be acted upon by the ATPG engine.

Fault Equivalence: The part of the ATPG process where the ATPG tool conducts fault management by declaring the equivalent faults to a fundamental fault and then removes equivalent faults from the fault list. Equivalent faults are those that are automatically detected when a fundamental fault is exercised—for example, detecting the stuck-at 0 fault on the output of an AND-gate requires driving the output to a logic 1, which requires driving all inputs to a logic 1, thereby detecting the stuck-at 0 faults on all inputs.

Fault Exercising: Logic values are placed on certain control points to create a portion of a vector that excites the fault. An ATPG tool will place a fault in a design description, and will trace back from the fault location to the necessary control points to create the portion of the vector that will excite or exercise the fault (drive it to the opposite of its faulted value for a stuck-at fault, or apply a logic transition for a delay fault).

Fault Grading: Also known as vector grading—the act of simulating a target vector against a good circuit description, and simultaneously a circuit description that contains a fault—the goal being to see whether the expected response is different between the two circuits at an observe point. If a difference is detected, then the fault has been detected—if a difference is not detected, then the fault is masked for that vector (not detected).

Fault Masking: A fault that is not able to be detected due to a circuit configuration problem such as reconvergent fanout or redundancy. An exercised fault cannot be driven uniquely to an observe point.

Fault Metric: The measurement of the fault content of a circuit—see *Fault Coverage*.

Fault Model: A mathematical model of faulty behavior that can be used to assess the compliance of a circuit to various criteria. For example, structural compliance can be verified by applying a fault based on a stuck-at fault model, timing compliance can be verified by using a fault based on a delay fault model, and current leakage compliance can be verified by using a fault based on a bridging fault model.

Fault Selection: The part of the ATPG process where the tool selects a fault from the total list of faults contained in the circuit.

Fault Simulation: see *Fault Grading*.

Faulty Circuit: A circuit description that contains a fault and is used for ATPG and fault simulation.

Firm Core: A core that is deployed as a gate-level model or design description more commonly called a gate-level netlist.

Flip-Flop: A sequential element comprised of a master-slave latch pair. There are many types of flip-flops (T, J-K, D, R-S), but the type mostly supported for scan testing is the D Flip-Flop.

Frequency Verification: Any test methodology used to verify the frequency of operation of a chip. The verification usually entails proving that register-to-register transfers can occur within one clock cycle when the maximum clock rates are applied. The three most common types of frequency verification are functional testing, scan-based delay testing, and static timing analysis signoff.

Full-Scan: A scan test architecture in which all sequential elements that are not specialized memory cells are scan cells connected into scan chains—and proper logic precautions have been taken to allow for the ability to safely shift random logic values through the scan chains.

Functional Scan-Out: A diagnostic technique where a scan architecture can be used to present the internal state of the device by scanning out the state of the internal registers during the application of functional vectors.

Functional Testing: A form of testing where a design element is tested by the application of functional, operational, or behavioral vectors. In general, these vectors are graded for structural coverage when applied to the final product testing. If the vectors are not structurally graded, then the testing should be called "design verification."

Gate Delay: The time required for a signal to propagate through a single gate from an input to an output.

Global Routing: Signal routing that is widely distributed at the chip level. In core-based design, a global route is a signal that is in the chip area outside the cores (even though the signal may be used within the core).

Good Circuit: A circuit description without any faults used as the comparison standard against which the faulty circuit is evaluated.

Hard Core: A core that is deployed as a layout model or design description (GDSII)—this type of core is also referred to as a layout macrocell or megacell.

HDL: *Hardware Description Language*—a text-based design description format based on very high-level software objects as compared to the use of gate and sequential objects with RTL.

IC: *Integrated Circuit*—the inclusion of many transistors and gate-level elements on a single silicon die.

Iddq Testing: Also known as quiescent current testing—a form of testing where an IC is placed in a state and after the combinational switching has settled, the current drawn by the IC is measured. In CMOS processes, good transistors draw no current other than diode reverse current when the transistor is not actively switching.

IEEE 1149.1: see *JTAG*.

Infant Mortality: The tendency for some silicon devices to fail early in their lifetimes due to latent defects. The failure statistics for silicon devices generally follow a "bathtub curve" where there is a group of failures at the beginning (infant mortality), a smaller number of failures during the normal operating lifetime, and then a rise in the number of failures at the "end of life."

Input Stimulus: The applied logic values to a circuit under test.

JTAG: *Joint Test Action Group*—the name of the group that began the IEEE 1149.1 Boundary Scan standard. It is now commonly used in slang form as a synonym for the 1149 set of standards.

JTAG Test Wrapper: A test wrapper that uses, and maintains compliance with, the boundary scan architecture, TAP, and TAP controller described in the IEEE 1149.1 Standard. A JTAG test wrapper uses a "slow scan" method that allows functional vectors to be serialized, scanned into the interface, and then applied "all at once" with an update operation. JTAG test wrappers generally do not support the needs of at-speed test or AC test goals. JTAG is mostly associated with the chip package pin interface, and not individual cores.

Known State: A requirement for conducting testing. A circuit under test must have a known state so that a deterministic (predictable) response can be generated by the application of a known input stimulus.

Latch: A state holding element that either can be transparent or can hold state. For scan testing, a latch-based design can be converted to the LSSD style of scan architecture.

LBIST: see *Logic BIST*.

Leakage Current: Unwanted current flow—in CMOS designs, when the transistors are not switching, there should be minimal current flow; similarly, when high-impedance devices are driving the logic Z state, there should be minimal current flow through the device.

LFSR: *Linear Feedback Shift Register*—a circuit device comprising a shift register and exclusive-OR gates providing feedback to some of the shift terms. The feedback terms are called the polynomial, and certain primitive polynomials will allow the shift register to take on all possible states except the all 0s state. The LFSR is used for BIST as either a pattern source (PRPG) or a response evaluation and compression unit (Signature Analyzer).

Load and Park: A scan test method where some scan chains are loaded and then parked (usually by stopping the clock to those scan chains), and then other scan chains are loaded before the sample cycle is applied. This type of testing is applied when there are timing concerns, or when independent scan architectures can't be operated simultaneously (for example, they borrow the same functional pins as the scan interface).

Logic BIST: The application of Built-In Self-Test to logic circuits—a form of testing for logic circuits where the stimulus generator and/or the response verifier is placed within the circuit. The most common form of logic BIST is to use linear feedback shift registers (LFSRs) to conduct pseudo-random pattern generation (PRPG) and to conduct output pattern compression (Signature Analysis). LBIST may be applied by converting functional flip-flops to the PRPG and Signature LFSRs, or adding separate PRPG and Signature LFSRs to drive and observe logic directly or to drive and observe scan architectures.

Logic Gates: Configurations of transistors that result in the simple logic functions of AND, OR, NOT, and the compound functions of Exclusive-OR, Exclusive-NOR, NAND, and NOR. In standard cell digital designs, many compound non-state-holding logic-gates representing complex logic functions are created (for example, an AND-OR-INVERT-gate), and the collection of these logic gates within a design is generally referred to as "combinational logic."

LSSD: *Level Sensitive Scan Design*—an architecture and method of scan design based on master and slave latches with separate phase clocks.

Macro-Cell: A completed design unit meant to be used "as delivered." For example, a macro-cell may be a hard layout block of a memory array meant to be placed on the floorplan intact, or it can be a soft core meant to be synthesized as a stand-alone object and placed on the floorplan intact.

Manufacturing Load Board: A backplane that connects a test platform (tester memory, data channels, power supplies, clock generators, etc.) to the integrated circuit's "chip socket." The load board should be designed to meet or exceed the limitations of the chip under test (frequency, power, channels, signal integrity, etc.).

Manufacturing Testing: Testing accomplished after the manufacture of a silicon product. The purpose of this testing is to verify that the manufacturing process did not add destructive defect content to the chip design. This type of testing is generally applied to packaged parts to ensure that the packaging process also did not add destructive defect content to the final product.

Mask: A set of photographic films used to make silicon dice on a wafer.

Masking: see *Fault Masking* and *Vector Masking*.

MBIST: see *Memory BIST*.

Measurement Criteria: What is being measured—in the language of test, a tester conducts an observation event and the observed data is compared to the "expected response"—if there is a mismatch, then a failure has occurred. The failure can be linked to "faults" and "defects." Even though the tester can discriminate only failures, the measurement statistic most cited is "fault coverage."

Mega-Cell: A large complex Macro-Cell. For example, a microprocessor core.

Memory BIST: *Memory Built-In-Self-Test*—the application of BIST to memory testing. Generally, this entails having an on-chip address generator, data generator, and read/write sequencing controller that applies a common memory test algorithm.

Memory Black-Boxing: see *Black-Boxing*.

Memory Testing: The application of test to memory arrays or memory-like devices. Memory testing is more "defect-based" rather than "fault-based," and memory defect diagnosis is conducted using "multiple clue" analysis.

Metal Layers: Transistor and gate elements are connected to each other by metal connections. Since the placement of elements and the routing of connections is not an easy problem to solve in just two dimensions (a single flat plane), then routing is generally handled by supporting multiple metal interconnect layers and having "vertical" connections called "vias" to enable connections to use the multiple metal layers.

MISR: *Multiple Input Signature Register*—an LFSR configured as a signature analyzer that allows multiple data input ports.

Modeling: The representation of a real physical device or physical effect with a mathematical description that can be manipulated and analyzed by a computer. For example, physical logic elements are made of layout polygons, and these can be modeled as transistors, and the collections of transistors can be modeled as logic gates—a design description made of logic gates can be operated upon by an ATPG tool to generate vectors.

Mostly Full-Scan: A partial-scan test architecture where most of the sequential elements, other than specialized memory cells, are scan cells and are connected into scan chains. The sequential cells that are not made scannable are those that are in areas of critical timing where scan insertion would negatively affect chip performance, or in large register arrays where scan insertion would negatively affect the chip area budget.

Multiple Clock Domains: A device that contains more than one clock distribution network. Different clock distribution networks are defined as separate physical distribution trees even if they are meant to operate at the same frequency.

Multiple Scan Chains: A device that contains multiple scan chains that may all be contained within the same scan architecture, or may be distributed across multiple scan domains. See also *Parallel Scan*.

Multiple Scan Domains: A device that contains multiple separate independently operated scan architectures that are associated with different clock domains, or are provided as separate independently operated scan architectures for some other reason.

Multiple Time Frames: When test pattern generation is conducted where multiple clock cycles are required to exercise a fault or to propagate a fault effect to an observe point, then the vector generation engine must calculate the state of the device for each clock interval—each clock interval is a "time frame" and can be treated independently by passing the states from surrounding time frames as constraints.

Multiplexor Mode: A method of connecting an embedded core, with or without a test wrapper, to the package pins of the part for access to test operations—this term is generally applied to an embedded core that does not support any other form of testing but the application of functional design vectors through the functional interface or through a bus interface brought out to the package pins.

Muxed-D Flip-Flop: A D Flip-Flop used for scan testing by placing a multiplexor in front of the D input port. The multiplexor adds the "scan data input" port, SDI, and is controlled by the "scan enable" port, SE.

Net: An implicit or explicit descriptor for a connection in a netlist. For example, a connection can be accomplished explicitly by creating a "wire" descriptor such as net332 and placing this descriptor as the output of a logical gate, across any hierarchical and/or interface boundaries, in any modules that it exists within and into the input or inputs of other logical gates; or a connection can be accomplished implicitly by placing the descriptor of an output connection associated with a gate output in the inputs of other gates.

Netlist: A text-based design description that represents the schematic of the circuit. A netlist can be switch-level (transistor) or gate-level. For example, a gate-level netlist describes a design by including the input signals, output signals, the instances of all gate-level sequential and combinational elements, and the nets that connect the gate-level elements to each other and to input and output signals.

Non-Core Logic: The logic in a chip design that is not delivered as a hard or non-mergeable core. This logic is generally named UDL (User Defined Logic).

Non-Mergeable Core: A core that may be delivered as a hard, soft, or firm core, but is delivered with DFT included (and usually with reuse vectors), where the "use" strategy is to not mix the logic with other chip-logic, and to reuse the test logic (and reuse vectors) as delivered.

Non-Recurring Cost: The "one-time" cost incurred on a silicon product. For example, it includes the cost-of-test that is added to the product during the design-for-test development and the vector generation process.

Non-Return Format: A tester waveform that can be described as a "print on change" data format, where the data applied to a pin by a tester changes only when the data in the pattern file changes. See also *Tester Edge Sets*.

Observability: The ability to observe nets, nodes, gates, or sequential elements after they have been driven to a known logic state. This has two main contexts, the ability for an ATPG tool to drive fault effects to observe points in the design description during the vector generation and the ability for a tester to observe expected response values of the chip internals at the package pins.

On-Chip: Any logic contained within a chip design. On-chip test logic is any logic within a chip design for the use of test; and on-chip logics that affect the test design are items such as on-chip clock sources (PLLs) and on-chip memory arrays.

Operational: A term that relates to the behaviors, functions, or operations of a chip design. See also *Functional Testing*.

Parallel Scan: The architectural support of multiple scan chains that are designed to be used simultaneously to reduce the number of shift clocks (shift bit depth) involved with the scan process. This is accomplished by parallelizing the scan bit depth across several tester channels. This architecture reduces the required clock cycles needed to load a state in the design unit and reduces one aspect of the cost-of-test.

Parametric Testing: A form of testing where the voltages, currents, and timing are measured for compliance against specified values. See also *AC Parametric Testing* and *DC Parametric Testing*.

Parking: see *Load and Park*.

Partial-Scan: A scan test architecture that allows sequential elements other than specialized memory cells to not be scan cells and to not be connected into existing scan chains. Usually only those cells that restrict the ability to achieve a high-fault coverage metric are made scannable and connected into scan chains.

Path Delay Fault Model: A mathematical model of faulty timing behavior involved with a complete described logic pathway being slow-to-rise or slow-to-fall. This model is used to verify the timing compliance of a circuit pathway. This model is applied by driving the path to the "fail" value initially, then by applying a logic transition to the "pass" value, and then observing the propagated fault effect at a later time interval.

Path Sensitization: The establishing of a propagation pathway from one point in a circuit to another point in a circuit to propagate a fault effect to an observe point. This analysis is done by an ATPG tool during the generation of vectors.

Pattern: A grouping of individual vectors. Usually the grouping of vectors has a purpose such as "the logic verification pattern," "the memory test pattern," or the "input leakage pattern." Sometimes vectors are grouped into small pattern sets to fit within the memory limits of a test platform, and several of these patterns collectively make up a purposeful pattern such as "the logic verification pattern set."

Pin Specifications: The timing specifications involved with the ability to apply new data to chip input pins and to observe valid chip output data. The specifications are usually defined as input-setup, input-hold, output-valid, and output-hold.

Potential Fault Detection: When ambiguity is involved in the observation of a fault at a legal detection point and the probability of detecting a fault is less than one, then the fault is classed as potentially detected. For example, the stuck-at fault applied on the enable pin of a tristate driver is generally classed potentially detected, since the 0-Z, 1-Z, Z-1, and Z-0 transitions are not well understood by many ATPG and fault simulation tools.

Primary Input: An input signal that is directly controllable by an ATPG tool during vector generation. If the vector generation is done at the chip level, the primary inputs are the chip package pins—if the vector generation is done at the core level, then the primary inputs would be the test access inputs (the functional inputs of the core would be set to logic X during isolated core ATPG since there is no UDL connected).

Primary Output: An output signal that is directly observable by an ATPG tool during vector generation. If the vector generation is done at the chip level, the primary outputs are the chip package pins—if the vector generation is done at the core level, then the primary outputs would be the test access outputs (the functional outputs of the core would be set to logic X during isolated core ATPG since there is no UDL connected).

PRPG: *Pseudo-Random Pattern Generation*—the generation of a repeatable random pattern sequence. For example, an LFSR can be used to generate a random-seeming sequence, and that sequence can be repeated if the LFSR is initialized with the same seed.

Qualification: A procedure by which produced silicon must meet compliance against a specification before it is allowed to be sampled, sold, or authorized to be submitted to full volume production.

Quality: A term associated with the overall goodness of a part. In the test industry, this term is taken as meaning the measurable fault coverage of a delivered device plus the device's reliability statistics.

Quiescent Current: In CMOS design, transistors consume power only during the switching event—when the transistor is in a stable state, the only current flow is that current associated with leakage involved with diode reverse currents. See also *Iddq Testing*.

Reliability: A term associated with the defect-free lifetime of a product (how long the product lasts in the field before it fails due to a latent defect or a degradation-over-time). The metric for reliability is generally a mean-time-between-failures (MTBF) rating or a failures-in-time (FIT) rating.

Recurring Cost: The cost that is incurred and repeated each time a product is manufactured. Part of recurring cost is the cost-of-test incurred with each device—the DFT logic impact on each die and the "in-socket" test time for each product that is tested.

Response Compression: The reduction of a circuit's output response to a much smaller amount of data. The reduction of a circuit's test data output stream can be done with a signature analysis LFSR such as a MISR.

Retention Testing: A form of testing to assess whether the state elements (sequential and memory) of a circuit are able to retain logic state for a long period of time. This form of testing is done routinely on memory arrays and is sometimes accomplished on sequential logic elements if the design is advertised as "static."

Robust Fault Detection: A term applied to delay testing where a path is exercised uniquely and where all off-path values required to enable the propagation path being tested are held stable for the number of time intervals required to conduct the test. For example, a two-cycle test that launches a 0->1 transition into a path must have all off-path logic values remain stable for the two cycles—if the off-path values toggle, then the resulting transition at the observe point may not be uniquely due to the launched transition.

RTL: *Register Transfer Level*—a text-based design description where the behavior of a circuit is modeled as data flow and control from register to register in reference to an applied clock, clocks, or other synchronizing signals.

Scan Cell: A sequential element connected as part of a scan chain—for example, a D flip-flop having a scan multiplexor to allow selection between the functional D input and the test-only SDI input. Scan flip-flops, when in functional mode, may support all specialty sequential functions such as Set, Reset/Clear, Data Hold, Clock Enable, and Asynchronous functions; however, the scan shift function must have higher priority over all other flip-flop functions.

Scan Chain: A set of scan cells connected into a shift register by connecting the scan cell's Q, or SDO, output port of one flip-flop to the dedicated SDI scan input port of another scan flip-flop.

Scan Design: The implementation of a scan architecture consisting of using scannable sequential elements, and including the scan data and control connections.

Scan Domain: A set of scannable flip-flops that are connected and synchronized by the same clock and are controlled by the same scan enable—for designs with multiple clock domains, multiple scan domains can be shifted and sampled independently of each other, and can be operated in such a way as to preserve the natural timing relationship between clock domains if necessary.

Scan Insertion: see *AC Scan Insertion* and *DC Scan Insertion*.

Scan Mode: The configuration of a design unit that supports the use of the scan chains for test purposes—more specifically, test control that allows applied scan data to be read into the first scan element of certain scan chains, allows test data to be read from the last element of certain scan chains, and allows the scan enable signal to be used to control the shift or sample action of certain scan chains—other internal control signals may also be asserted. A signal named "scan mode" is sometimes created to constrain certain pins and circuit elements during scan testing.

Scan Path: Another term for pathways created by the scan chains—the shift paths evident in a design with a scan architecture.

Scan Test Wrapper: A design unit or level of hierarchy that is viewed as the external interface of a non-mergeable or "test-in-isolation" core, and that uses scan chains routed through the interface hierarchy level and/or test access to internal scan chains as the method to apply and evaluate isolated embedded testing.

SDI: The *Scan Data Input* port on a scan cell—more specifically, the assert side of the input multiplexor when the control signal, SE, is asserted.

SDO: The dedicated *Scan Data Output* port on a scan cell, or the connection on the scan cell Q or QB output port used as a dedicated scan connection to another scan cell's SDI port.

SE: The *Scan Enable* port on a scan cell and its globally routed control signal—this is the multiplexor control signal that selects whether the scan cell will update with data from the functional D input port or the test-only SDI input port—this signal is generally distributed as a fanout tree, since it must control every scannable flip-flop in a design.

SE—Fast_SE, Slow_SE, Shift_SE: When multiple clock domains exist, and if clock skew versus data is not managed sufficiently, multiple scan enable type signals, SE, must exist to provide independent shift and sample control of portions of the design and to create the independent scan domains—for example, even though there is no required naming convention, a Fast_SE could be the scan enable (SE) signal distributed to all scannable flip-flops connected to the fast clock, while a PLL_SE could represent the scan enable signal distributed to the scan chains in the Phased-Lock-Loop design element.

SE—Force_SE, Tristate_SE: When scan design rules are relaxed on asynchronous elements, gated-clocking, and tristate or multiple driver nets, then scan control signals are needed to ensure that safe shifting can occur (scan data is not stopped or corrupted during the shift operation, and driven contention does not occur during the shift operation). These signals may not fanout to every scan flip-flop, and so they should be modeled in the HDL/RTL and synthesized to make timing as opposed to SE signals, which are gate-level scan inserted.

Seed: The initial state of an LFSR. LFSRs can be "seeded" to provide different pseudo-random sequences by installing different seeds—this process is known as re-seeding.

Sequential: A type of circuit or circuit description that is comprised of both combinational gates and state holding or sequential logic in the circuit description. Circuits that are sequential require multiple time frame combinational ATPG or sequential ATPG to process.

Serendipitous: Bonus fault coverage that is discovered when a vector for a targeted fault is fault simulated. For example, a vector is created to target only one fault; when this vector is fault simulated, it is discovered that 30 other faults were detected—the 30 extra faults are serendipitous fault coverage. Another form of serendipitous fault coverage is when vectors generated against one fault model are graded against another fault model—for example, a set of vectors with 97% stuck-at fault coverage is fault-simulated against the transition delay fault model, and 35% serendipitous transition delay fault coverage is discovered.

Shared Pins, Borrowed Pins: Package pins that have extra uses other than their functional purposes—more specifically, pins that are used to provide direct access to scan chains or direct access to embedded core units when those units are in a test mode.

Signature: The final value left in an LFSR after a test sequence has completed. The changing values in the LFSR during the test sequence may be referred to as incremental signatures if they are accessed for diagnostic purposes.

Signature Analysis: The process of compressing an output response into an LFSR. If the final signature in the register is identical to an expected signature, then the test has passed—if there is a difference, then the test has failed.

Signature Dictionary: When a signature analyzer is used for output response compression, a method of diagnosis can be supported if a table or dictionary of "failing" signatures is kept—"this failure results in this signature." Note that most signature dictionaries are based on the single fault assumption.

Silicon Die: A single silicon device. Silicon die are the individual silicon devices separated by scribe grids on a wafer, or loose die are the die after they have been "sliced-and-diced" and removed from the wafer.

Simulator: A software device used to evaluate the response of a circuit to some applied stimulus. For example, gate-level simulation would have ATPG vectors applied to gate-level models to evaluate the circuit's response.

Single Fault Assumption: The assumption that an ATPG engine or a fault simulator uses to simplify its processes. It is much easier to conduct the various analyses with only one fault installed in the design description—including the ability to conduct analyses with multiple faults significantly increases the compute resources needed.

Slow-to-Rise, Slow-to-Fall: A description of the fault effect attributed to the delay fault models.

Static: A device description that represents the ability to operate the device at zero Hertz without data loss or destructive results. In most cases, this means that no "dynamic" or active logic is supported (such as the refresh required by DRAM or the pre-charging of tristatable busses). The term *static* is also sometimes applied to "static for low power," which adds the extra requirement of eliminating or disabling current drain devices such as pullup and pulldown logic on busses and package pins.

Standard Cell: A standard logic or sequential element in a "library" of elements available for design use. Generally, the library cells are pre-made and are associated with specific silicon processes that are then "mapped" to the design through a synthesis process.

Standard Cell Design: A design style that uses standard cells as opposed to designing directly at the transistor level (full-custom and semi-custom design).

Structural Testing: A form of testing whereby the goal is to verify the structure of a chip (the wire connections and the gate truth tables). The opposing form of testing is functional or behavioral testing.

Stuck-At Fault Model: A mathematical representation of faulty behavior based on shorting gate connections and wire routes to either VDD or VSS. The SAF is a DC fault model that is applied independent of timing or frequency.

TAP: *Test Access Port*—a set of four standard pins (TMS, TDI, TCK, and TDO) and one optional pin (TRSTB), as defined by the IEEE 1149.1 standard to enable boundary scan testing.

Test: The observation of a known expected response as a result of the application of a known input vector into a circuit in a known state—the purpose being to measure some response against a compliance standard.

Testability: see *Design-for-Testability*.

Test Coverage: A metric that is slightly different from fault coverage. Test coverage can be defined as the cumulative fault coverage of the testable logic (non-testable or redundant faults are removed from the denominator). Test coverage is generally higher than fault coverage.

Tester Accuracy: The smallest ambiguity in edge placement.

Tester Channel: Generally, the memory behind one tester pin; however, sometimes testing techniques connect more than one tester pin/channel to a single chip pin.

Tester Edge Sets: A tester timing format that can be applied to each pin individually. An edge set, for example, is the data format and timing information involved with applying new data to a chip input pin and is comprised of the input-setup time, the input-hold time, and the waveform type such as non-return (NR) or return-to-zero (RTZ). A chip package pin may support more than one edge set during testing.

Tester Precision: The smallest measurement that the tester can resolve.

Tester Timing Format: A term related to tester vector formats. Testers may contain waveform generators that make use of various waveform types such as non-return (NR), return-to-zero (RZ), and surround-by-complement (SBC), which are designed to change logic values at the setup and hold times associated with an input pin. A test pattern comprised of logic 1s and 0s can be mapped to these various formats during the development of the test program. These timing formats are designed to verify the device pin specifications with each vector application.

Test Logic: The logic included in a chip design to implement specific test functions. For example, the logic associated with a scan test architecture.

Test Program: A collection of test patterns that has been organized in a certain order and converted to the language (data format) of a specific test platform. It is called a program because it is basically software, and the tester is the CPU—test patterns may be comprised of subroutines, pattern and test routine calls, and algorithmically controlled or sequenced events.

Test Wrapper: A design unit or level of hierarchy that is viewed as the external interface of a non-mergeable or "test-in-isolation" core, which also contains logic that allows isolated testing to occur—the testing supported can be BIST, Scan, or direct access from multiplexed pins. A test wrapper is usually designed to reduce the functional interface of a core to a more manageable number of test signals to ease test integration.

Test Vectors: The logic 1s and 0s applied to a device specifically for the test process (as opposed to vectors used for operational purposes).

Time-to-Market: *TTM*—the amount of time it takes to get from a design specification to a product sample. Time-to-market generally concerns the design aspect of a product development cycle.

Time-to-Volume: *TTV*—the amount of time it takes to get a chip into volume production. Time-to-volume generally concerns the vector delivery, tester management, and yield enhancement aspects of the development cycle.

Toggle Fault Model: A mathematical model that represents the behavior of driving every net, node, gate input, and gate output to both logic states, but not associating a propagating observation event (just driving each element to both states is sufficient; propagating the effect to a valid observe point is not necessary).

Transition Delay Fault Model: A mathematical model of faulty behavior that is based on a gate or gate connection being slow-to-rise or slow-to-fall and is used to verify the timing compliance of that gate or gate connection. This is accomplished by driving the fault to the fail value initially, then by applying a logic transition, and by observing the propagated fault effect at a later time interval.

UDL: *User Defined Logic*—in core-based design, the additional logic added to the chip design that is not a hard or non-mergeable core. See also *Non-Core Logic*.

ULSI: *Ultra Large Scale Integration*.

Undetected Faults: Faults that are not exercised or observed by the applied set of test vectors.

Vector: The collection of logical 1s and 0s applied to a chip at a given point in time. Usually the vector is referenced from a synchronizing clock signal—in this case, the definition of a vector is the collection of logical 1s and 0s applied per clock edge. Scan vectors are a special case where a single scan vector is the collection of logical 1s, 0s, and the number of clocks required to load a scan chain.

Vector Compaction: The reduction of the applied test vector data by placing many separate tests into few vectors or by removing redundant vectors from a pattern set so that only the most efficient vectors remain.

Vector Compression: see *Response Compression*.

Vector Efficiency: Also known as "fault efficiency" or "fault efficient vectors"—a metric that concerns the "faults per clock-cycle," or the fault coverage content of a vector set in relation to the size of the vector set. A rating of more faults-per-vector is generally good, but a higher rating indicates vectors with high activity content that may consume more power during application.

Vector Masking: The ability to ignore an "expected response value" of an output vector. This has two applications, the ability of an ATPG tool to print vectors in a data format that includes a vector mask (a pseudo-vector that maps onto an output vector and indicates which bits of the vector are "cared") and the ability of the test platform to ignore output values that are not required as part of the test on a vector-by-vector basis.

Verification: The process of proving that a design is in compliance with a specification. In most cases, the specification is a data sheet or data manual, but sometimes the specification is the HDL or RTL model, and the verification is proving that the logic produced by the synthesis process, scan insertion, or the place-and-route process is boolean and operationally equivalent.

Verilog: A popular text language for HDL, RTL, and netlist design descriptions.

VHDL: A popular text language for HDL, RTL, and netlist design descriptions.

Via: A vertical metal connection between metal interconnect layers. See also *Metal Layers*.

VLSI: *Very Large Scale Integration*.

Wafer: A large round silicon substrate that provides the base upon which the individual silicon dice will be manufactured.

Wafer Probe: The application of some level of testing to dice prior to the dicing process to remove them from the wafer.

Winnowing: Eliminating redundant vectors from a vector set to reduce or optimize the size of the test data file.

Worst Case: A condition where all variables in a process are assumed to be at the "worst" limit. For example, if gates in a circuit are all rated at a propagation timing with a plus or minus variation, then worst case is the condition of all delay being at the "plus" end so that the timing goals are missed by the most amount. Worst case, in silicon design, is generally applied to the process variables involved with fabrication.

X: A conceptual state or pseudo-logic value used to represent the "don't care," "don't know," or "unknown" values associated with circuit inputs, outputs, and states. These values may be evident in ATPG, simulation, or in the test data applied to a tester.

Yield: In an ideal sense, yield is a metric representing the number of "good dice" on a wafer compared to the total number of dice manufactured on that wafer. However, yield depends on the number of defects present on the wafer, the process variation that would cause out-of-compliance behavior, and the "goodness" of the test program. (Does the test program allow defective and faulty parts to escape? Or are the measurement criteria so strict that very few parts pass?)

Z: A pseudo-logic value used to represent the "high impedance," "undriven," or "midband" values associated with tristate nets, input-output, and output signals. These values may be evident in ATPG, simulation, or in the test data applied to a tester.

Index

Numerics
$2^{(n+m)}$ testing 21, 99
2^n testing 20, 99
2^n-1 states 168, 169

A
abort limit 62
AC
    fault model 17
    frequency assessment 30, 31
    logic delay 30, 31
    pin specification 30, 31, 307
    scan 146, 148, 258, 286, 300
    scan testing 145, 148, 158
    testing 145
access time 186
accuracy 25, 27, 28
address 183
address bus 183
algorithm 177, 218, 220
aliasing 170, 232
analog functions 32
application specific integrated circuits *see* ASICs
apply data 44
ASICs 42, 87, 246
aspect ratio 186
ATE 23, 25, 191
ATPG 3, 23, 37, 38, 44, 62
ATPG process 38, 41, 44, 45
at-speed scan 148, 150, 152, 266
at-speed testing 148
automated vector generation process 38
automatic test equipment *see* ATE
automatic test pattern generation *see* ATPG

B
backtrace limit 62
backtracing 44, 63
balanced scan chains 125, 139, 244
banked MBIST 229
benchmarking 87–90, 91
BIST 21, 168, 169, 206, 208, 218, 220, 222, 234, 236, 237, 240, 244, 252, 266, 302, 309
BIST controller 225
BISTed memory 225
bitmap 287
bitmap data 220
bitmapping 240
bitmapping diagnosis 203
black-boxing 150, 208, 212
borrowed functional pins 128
brown out 229
built-in DFT 246
built-in self-test *see* BIST

C

cared data 142
channels 25
characterization 21, 49, 259
checkerboard pattern 201
circuit learning 56, 91
clock
    data race 154
    domain 49, 154
    skew 103, 140, 154, 157, 158, 173, 252, 305
    source 131, 173, 281, 292
    versus data 140
clocked-scan 104
clocked-scan cell style 103
clock-in reference 27, 131
clock-out reference 27, 131
clock-to-out 273, 305
clock-to-Q 50, 103, 140, 157
CMOS 13, 18, 47, 52, 180
combinational ATPG 73, 84, 101, 108
combinational stuck-at fault 47
combinational-only 47, 73
combinational-only ATPG 173
combinational-only circuit 101
comparator 231, 234
complementary metal oxide semiconductor *see* CMOS
concurrent engineering 4, 91
concurrent fault simulation 77
configuration management 80
contention checking 120
contention-free 173
contention-free vectors 120, 134, 135
control point 44, 101
controllability 10, 44, 54, 55, 56, 177
core-based
    design 41, 243, 246, 248, 253, 256, 311
    design methodology 311
cores 4
    deliverables 250
    power 254
cost-of-test 3, 4, 8, 25, 27, 32, 37, 42, 91, 95, 96, 125, 145, 177, 180, 186, 243, 244, 298, 311
COT *see* cost-of-test
critical paths 69, 145, 146, 163, 164, 165, 166
critical timing paths 69
current leakage 18, 52, 71
current measurement 18, 25, 71
current-based fault model 18
current-based testing 11, 45, 52, 71, 84
cycle-based ATPG 75

D

data
    apply 44
    bus 183, 218
    expect 44
    format 37, 44
    mask 44
    smearing 157
    vector 44
DC
    logic retention 30, 31
    logic stuck-at 30, 31
    pin parametrics 30
    scan 128, 148, 154
    scan insertion 139, 161
    testing 145
debug diagnosis 220, 240
dedicated test pins 262
defects 13, 16, 18, 47, 142, 143
delay defect 50, 145
delay fault 13, 49, 50, 145, 166
design for reuse 256, 257–260
design rule checks *see* DRC
design rules 39, 41, 42, 83–85, 91, 134, 134–137, 173
design verification 20, 30, 99
design-for-test *see* DFT
detect points 23, 63
detection 47
determinism 10
deterministic detection 47
deterministic vector generation 80
DFT 3, 7, 134, 173, 246, 248, 252, 253, 256, 311
diagnosis 52, 71, 170, 231
diagnostic 41, 80, 166, 170, 173, 203, 231
diagnostic fault simulation 142, 143, 166
diagnostic fault simulator 142, 143
Direct Memory Access *see* DMA
direct test access 265

Index

distributed memory arrays 188
DMA 190
DNR 28
don't care 33
do-not-return *see* DNR
doubly embedded 177, 302
DRAM 180, 184, 186, 201, 237, 240
DRAM refresh 201
DRC 38, 103
driven contention 44, 84
drudgery 41
Dynamic Random Access Memory *see* DRAM

E

edge search 27
edge set 27, 298
edge-rate 25, 49, 131, 132
EEPROM 4, 180, 184, 186, 240, 266, 287
Electrically-Erasable Programmable Read-Only Memory *see* EEPROM
embedded core 243
embedded memories 188
exact paths 69
excitation path 44
exercise 47
exhaustive testing 20, 21
expect data 44
expected response 10, 77, 78

F

failure 13, 18, 47, 50
failure analysis 8
failure mode 13
failure modes and effects analysis *see* FMEA
fake word 208, 216
false paths 75, 140, 158, 165, 166
fault 13, 18
    AC model 17
    coverage 14, 20, 23, 41, 44, 66, 68, 78, 83, 91, 95, 99, 165, 166, 173, 244, 309
    current-based model 18
    definition 13
    delay 18
    detection 20, 44, 47, 68, 77
    dropping 45, 81
    efficiency 81
    enumeration 44, 63, 65
    equivalency 60, 63, 91
    exercising 63, 65, 66
    gate delay 65
    gate-level 16, 17
    grading 4, 14, 41, 71, 77
    masking 58
    metric 13, 14
    models 16, 37, 47, 54
    path delay 16, 37
    propagation 65
    selection 62, 63, 65, 68
    simulation 4, 44, 77, 80, 81, 83
    simulator 23, 41, 81
    single-stuck-at 16, 17
    stuck-at 14, 17, 37, 54
    switch-level 16
    toggle model 18
    transition delay 16, 37, 65
firm cores 252, 254, 293, 309
flash memory 4, 180, 184, 186, 240
FMEA 54, 177
forward tracing 44, 63
free-running clock 83
frequency verification 145
full-scan 73, 74, 81, 108
functional scan-out 142, 143
functional simulation 143
functional testing 20, 21, 97, 99
functional vectors 41, 99, 143, 146, 262

G

gate delay 17, 49, 65
gated clocks 84, 85
gate-level 16, 17, 47, 83, 163
golden standard 236
gross delay fault 66, 68

H

half-cycle paths 75
hard cores 244, 248, 252, 253, 254, 256, 265, 286, 293, 309
hard to detect 80
head registers 150, 152
heads 152, 161
histogram 164
hold time 66

I

Idd 30, 32, 45, 52, 309
Iddq 30, 32, 45, 52, 131, 132, 250, 252, 259, 281, 287, 300, 307, 309, 311
IEEE 1149.1 45, 220, 248, 265, 273, 292
Iih 30
Iil 30
independent shift bits 271
indeterminant 10
inductive testing 203
input
    hold 145
    hold time 28
    setup 145, 250, 273, 287, 298
    setup time 27, 50
    stimulus 11, 27
    stimulus, known 10
intellectual property *see* IP
inverse LFSR 231
Ioh 30
Iol 30
IP 4, 246, 253, 256
Issq 52
Iz 30

J

JTAG 45, 220, 246, 248, 258, 273, 275, 292

K

known input stimulus 10
known response 10
known state 10, 77, 118
known stimulus 77

L

last shift 112, 115, 148, 154
LBIST 168, 169, 257, 259
leakage threshold value 71
LFSR 168, 169, 170, 171, 222, 231, 232, 236, 266
library of standard cells 38
linear feedback shift register *see* LFSR
load board 23, 25
logic BIST *see* LBIST
logic X 11
logic Z 11
LSSD 104, 244, 253, 309
LSSD scan style 95, 103

M

macrocells 4
manufacturing test 8, 23, 78
March test 205, 220, 222, 237
mask data 44
massive observability 71
maximum power 52
MBIST 177, 190, 191, 218, 220, 222, 227, 229, 237, 259, 262, 286, 287, 302
MBIST comparator 223
MBIST interface 222
measurables 87, 91
megacells 4
memory
    addresses 218
    bridge 199
    bridging fault 197
    built-in self-test 177
    decode fault 199
    delay fault 199
    failure modes 193, 205
    fault model 193
    modeling 210
    power rating 186, 188, 225, 227, 229, 240
    retention 32, 307
    retention test 220
    stuck-at 199
    test algorithm 205, 206, 210, 220, 237, 240
    test function *see* MTF
    testing 31, 307
    transparency 208, 214
mergeable cores 252, 253, 293, 309
MISR 168, 236, 237
MISR LFSR 236
Moore's Law 41
mostly full-scan 122
mostly-scan 73, 74
MTF 191, 205
multiple clue analysis 177, 193, 203, 231, 240
Multiple Input Signature Analyzer Register *see* MISR
multiple scan chains 125
multiple time frame analysis 74
multiple time frames 73, 74, 75, 81
Mux-D scan 103, 104, 244
Mux-D scan style 95, 244, 253, 309

Index

N
NDA
nearest neighbor 108, 152, 157
non-defaulting cores 279
non-disclosure agreement *see* NDA
non-mergeable cores 244, 248, 252, 253, 256, 265, 293, 309
non-mergeable hard cores 252
non-mergeable soft cores 252
non-return format 28
non-return-to-one *see* NR1
non-return-to-zero *see* NRZ
non-robust fault detection 68
NR1 28
NRZ 28

O
observability 10, 54, 55, 56, 177
observe point 44, 47, 49, 52, 63, 65, 77, 101
off-path values 68
output
    hold 145
    hold time 28
    response 27
    valid 145, 250, 273, 287, 298
    valid time 28

P
P1500 Standard 246
package cost 289
parallel scan 277, 286
partial scan 73, 74, 108, 122, 123, 132
path 17
path delay 17, 37, 45, 49, 68, 74, 309
path delay fault 16, 68
path descriptions 69
path enumeration 69
path filtering 165
path selection 68, 165
path selection methodology 163
path sensitization 63, 65, 66
patterns 11, 30
pin
    parametrics 307
    specifications 27, 145, 148, 150, 152, 164
    timing 4, 27, 28
    timing specification 145

porcupine 256
precision 25, 27, 28
primitive polynomial 169, 236
process characterization 8
propagation delay 49
propagation path 44
PRPG 168, 169
PRPG LFSR 168, 169, 170, 231
pseudo-exhaustive testing 20, 21
pseudo-random pattern generator *see* PRPG
pseudo-stuck-at 16, 18, 45, 52, 71, 300

Q
qualify 7, 41
quality 4, 41, 95, 243
quality exceptions 99
quiescent current 18, 287
quiescent leakage current 52
quiet-vector-selection 18, 45, 52, 71, 300

R
R1 28
RAM 287
random pattern resistance 168, 169
random pattern resistant faults 169
random vectors 80
randomness 10
RC 28
Read-Only Memory *see* ROM
redundant vectors 81
re-fault simulate 81
reliability 4, 41
response 11, 13
retention 201, 220, 223, 307, 311
retention testing 240
return format 28
return-to-complement *see* RC
return-to-one *see* R1
return-to-zero *see* RZ
reusable cores 311
reuse 4, 244
reuse cores 244, 246, 289
reuse vectors 80, 243, 244, 250, 257, 293, 296, 298, 304, 305, 311
    format 304
    frequency 304
    power 304

sizing 304
robust fault detection 68
ROM 180, 181, 184, 236, 237, 240, 266, 287
ROM MISR 237
row decoding 183
RZ 28

S
safe sampling 135
safe shift 106, 120, 139, 148, 150
safe shifting 118, 134, 135, 161, 173
safe vectors 83
sample skew 158
SBC 28, 201
scan 244, 252, 309
    cell 134, 173
    cell substitution 139, 161
    chain 108, 173, 258
    chain optimization 139, 161
    design misconceptions 96
    diagnostics 142
    domain 154, 173
    effective circuit 101
    flip-flop 101
    insertion 97, 101, 103, 148, 161, 271
    interface 128
    sample 110, 112, 115, 140, 148, 150
    shift 110, 112, 115
    shift data races 157
    test timing 115
    testing 21, 96, 134
    testing methodology 96
    testing misconceptions 97
scan-based testing 96
scan-out 143
SCOAP 55
SDI 103, 108, 110, 112, 128, 139, 157
SDI routing 139
SDI-SDO 157
SDO 103, 108, 110, 139, 157
SE 103, 106, 108, 110, 112, 128, 129, 161
SE routing 139
seed 169, 170
self-defaulting cores 279
sequential ATPG 73, 122, 123, 256
Serial Input Signature Analyzer Register *see* SISR

setup time 66
shift contention 148
shift data corruption 106
shift race 157
shift skew 157, 158
signature analysis 168, 169, 231, 238
signature analyzers 168, 170, 234
signature dictionary 170, 231, 232
simultaneous sample 158
single edge set 260
single-fault assumption 16, 52, 71, 142
single-stuck-at 13, 16, 17, 74
SISR 168
slack 163
slow-to-fall 18, 49, 65
slow-to-rise 18, 49, 65
SOC 3, 41, 99, 243, 244, 246, 248, 254, 292, 311
soft cores 252, 254, 256, 293, 309
specification compliance 244
specifications 27
SRAM 180, 184, 186, 240, 266
stable sample 159
Static Random Access Memory *see* SRAM
static timing analysis 163
stimulus 11, 13
stop-on-first-fail 30
stored pattern test 266
structural testing 20, 21, 47, 49, 173
structural vector 44
structural verification 99
structured methodology 96
stuck-at 14, 17, 37, 45, 47, 54, 71, 309
stuck-at memory bit cell 195
subtle delay fault 68
surround-by-complement *see* SBC
switch-level 13, 16
system-on-a-chip *see* SOC
system-on-silicon 16

T
tail registers 150, 152
tails 152, 161
test
    access 96, 177, 246, 265, 292, 311
    data rate 131, 173, 263
    data volume 99

Index

DMA 208, 225
DMA direct access 190
escapes 32
exhaustive 20
functional 21
in isolation 246, 259, 269, 275, 293, 294, 296
logic 7, 8
manufacturing 8
power consumption 260
program 23, 30, 33, 99, 250
pseudo-exhaustive 20
scan 21
structural 21
wrapper 250, 258, 268, 269, 271, 273, 275, 277, 286, 294, 298
testability 54
tester
  accuracy 28
  cost 289
  data formats 28
  format 27
  precision 28
tester-related cost 289
time-to-market 3, 4, 37, 41, 42, 91, 95, 123, 177, 243, 244, 248, 260, 289, 292, 311
time-to-volume 3, 4, 37, 42, 91, 99, 292
timing analysis 69, 271
timing format 27
timing histogram 164
toggle 45, 52, 71, 300
toggle fault model 18
transition delay 16, 17, 37, 45, 65, 66, 68, 74, 166, 309
tristate-valid 28, 145
TTM *see* time-to-market
TTV *see* time-to-volume
two-cycle analysis 74
two-time-frame testing 68, 75

U

UDL 248, 262, 269, 271, 275, 281, 292, 298, 300, 302
UDL vectors 298
ULSI 243
ultra-large scale integration *see* ULSI
user-defined chip logic *see* UDL

V

vector
  compression 260
  data 4, 25, 44
  data reduction 33, 39
  data sizing 33
  definition 11
  efficiency 4, 41, 42, 91, 146
  formats 80, 88
  management 260
  pair 17, 65, 74
  reload 25, 37
  set 39
  sizing 88
  structural 44
  translation 39
  volume 4, 32, 123, 173
vector pair 148, 150, 199, 273, 287
Vih 30
Vil 30
Virtual 269
virtual test socket 246, 269, 271
Voh 30
Vol 30
voltage-based testing 11, 45, 52, 71
VSIA 246

W

wafer probe 8
wired logic 84

X

X 10, 11, 25, 33, 81, 83, 85, 118, 120, 122, 126, 152, 154, 159, 170, 287
X-filled 126
X-filler 126
X-management 170
X-space 81

Y

yield enhancement 8
yield turnaround 142

Z

Z 11, 16, 25, 83, 85

About the Author

Al Crouch began his career in test and testability in the U.S. Air Force as a Meteorological Equipment and Weather Radar repairman. He went from the Air Force to the University of Kentucky, where he earned a BSEE, and continued on to get his MSEE by conducting a course of study with a focus on Quantum Electronic Devices. Since graduating, he has worked for Texas Instruments, Digital Equipment Corporation, and Motorola, always with a focus on design-for-test, test automation, and computer-aided test.

Al currently works for the ColdFire Core Technology Center within Motorola, and is a member of the technical staff. His field of work includes researching and implementing test technologies such as scan, at-speed scan, scan-based timing assessment with critical path selection, scan insertion, ATPG, logic BIST, memory BIST, JTAG, testable reuse cores, core test integration, automatic vector diagnostics, and vector data management. He has worked on such products as the MCF68060 microprocessor family, the ColdFire microprocessor family, the MCore, Star12, and several of the FlexCore ASIC+Core products. Al has published extensively in several IEEE publications, EE-Times, and industry journals, and he has been a frequent presenter at IEEE conferences, workshops, and symposia such as the International Test Conference, the International Test Synthesis Workshop, and the VLSI Test Symposium.

Al has had nine U.S. Patents issued on test-related inventions concerning Logic BIST, Memory BIST, Scan Architectures, Scan Optimization, Low Power Test, and At-Speed Scan. He is also a member of the IEEE.

Al can be contacted at *Al_Crouch@prodigy.net*.